EVOLUTIONARY THEORY

Other interview books from Automatic Press ♦ ∀P

Formal Philosophy
edited by Vincent F. Hendricks & John Symons
November 2005

Masses of Formal Philosophy
edited by Vincent F. Hendricks & John Symons
October 2006

Political Questions: 5 Questions for Political Philosophers
edited by Morten Ebbe Juul Nielsen
December 2006

Philosophy of Technology: 5 Questions
edited by Jan-Kyrre Berg Olsen & Evan Selinger
February 2007

Game Theory: 5 Questions
edited by Vincent F. Hendricks & Pelle Guldborg Hansen
April 2007

Philosophy of Mathematics: 5 Questions
edited by Vincent F. Hendricks & Hannes Leitgeb
January 2008

Philosophy of Computing and Information: 5 Questions
edited by Luciano Floridi
Sepetmber 2008

Philosophy of the Social Sciences: 5 Questions
edited by Diego Ríos & Christoph Schmidt-Petri
September 2008

Epistemology: 5 Questions
edited by Vincent F. Hendricks & Duncan Pritchard
September 2008

Complexity: 5 Questions
Carlos Gershenson
November 2008

Mind and Consciousness: 5 Questions
edited by Patrick Grim
January 2009

See all published and forthcoming books in the 5 Questions series at
www.vince-inc.com/automatic.html

EVOLUTIONARY THEORY
5 QUESTIONS

edited by

Gry Oftedal, Jan Kyrre Berg O. Friis
Peter Rossel and Michael Slott Norup

Automatic Press ♦ $\frac{V}{I}$P

Automatic Press ♦ $\frac{V}{|}$P

Information on this title: www.vince-inc.com/automatic.html

© Automatic Press / VIP 2009

This publication is in copyright. Subject to statuary exception
and to the provisions of relevant collective licensing agreements,
no reproduction of any part may take place without
the written permission of the publisher.

First published 2009

Printed in the United States of America
and the United Kingdom

ISBN-10   87-92130-26-7   paperback
ISBN-13   978-87-92130-26-6   paperback

The publisher has no responsibilities for
the persistence or accuracy of URLs for external or
third party Internet Web sites referred to in this publication
and does not guarantee that any content on such
Web sites is, or will remain, accurate or appropriate.

Typeset in LaTeX2_ε
Graphic design by Vincent F. Hendricks

Contents

Preface	v
Acknowledgements	vii
1 Patrick Bateson	1
2 John Tyler Bonner	5
3 Terrence W. Deacon	11
4 Daniel C. Dennett	17
5 Douglas J. Futuyma	23
6 Peter Godfrey-Smith	37
7 Brian Goodwin	45
8 David L. Hull	49
9 Eva Jablonka	63
10 Philip Kitcher	79
11 U. Kutschera	95
12 Richard Levins	103
13 Elisabeth A. Lloyd	109
14 Stuart A. Newman	125
15 Samir Okasha	143
16 Susan Oyama	153
17 David C. Queller	171

18 Michael Ruse	187
19 Geerat J. Vermeij	209
20 Andreas Wagner	223
21 David Sloan Wilson	231
Index	241

In the memory of Brian Goodwin,

It is with great sadness that we have received news of Brian Goodwin's passing away on July 15th 2009 – shortly after giving what has to be one of his very last interviews – which we are here honoured to publish.

Preface

Evolutionary biology is both a highly developed biological discipline and a scientific backdrop against which many other fields of research develop. The famous quote "Nothing in biology makes sense except in the light of evolution" from the molecular and evolutionary biologist Theodore Dobzhansky, was a reaction to anti-evolutionary creationism in the early 1970s. Evolution still is, perhaps more than ever, in the centre of influential discussions, reaching far beyond questions internal to the sciences.

This volume is published 150 years after "On the Origin of Species" by Charles Darwin. It is an appropriate time to have outstanding scholars in the field of evolutionary biology and philosophy presenting their views on the broader standing of biological evolution today. Both philosophers and natural scientists give in this book stimulating and thought provoking contributions on evolutionary research in an accessible manner. They answer questions about how they came to work on evolution, about their own research, about practical, political or moral implications of evolutionary science, and about interesting criticisms and future prospects.

The science of evolution is a field of research where natural scientists and philosophers work side by side developing models and theories. This exciting dynamic is reflected in this volume, and many different angles on the modern discussion are present. The contributors convey their messages in a way that will work as a resource both for students, researchers and a broader audience interested in the foundation, application and scope of biological evolution.

November 2009
Gry Oftedal, Jan Kyrre Berg Olsen Friis,
Peter Rossel and Michael Slott Norup
Editors

Acknowledgements

We are particularly grateful to the contributors for devoting time to writing such erudite, enlightening and often thought-provoking interviews and grateful to the philosophical community in general for showing interest in this project. In addition we would like to express our gratitude to editor-in-chief Vincent F. Hendricks and managing editor Rasmus Rendsvig and thus to our publisher Automatic Press ♦ $\frac{V}{I}$P, in particular senior publishing editor V.J. Menshy, for continuing to take on these 'rather unusual academic' projects.

November 2009
Gry Oftedal, Jan Kyrre Berg Olsen Friis,
Peter Rossel and Michael Slott Norup
Editors

1
Patrick Bateson

Emeritus Professor of Ethology

Sub-Department of Animal Behaviour, University of Cambridge, UK

---

1. Why were you initially drawn to discussions and research on evolution (or evolutionary aspects of your field)?

From a very early age, I told anybody who asked that I wanted to be a "biologist" without having any clear idea what that might entail. The reason was that I had a kinsman who was a very eminent scientist. His name was William Bateson and, as one of the champions of Gregor Mendel, he coined the term "genetics". He was a cousin of my grandfather and had died 12 years before I was born, but the family was evidently very proud of him and often referred to him. Many years later in my life I wrote an assessment of him and discovered how far ahead of his time he had been (Bateson, 2002).

As a schoolboy I was much taken by a double biography of Charles Darwin and his so-called bulldog, Thomas Henry Huxley (Irvine, 1955). This was my first taste of how important Darwin had been to modern biology.

Although my subsequent research interests became focused on ontogeny, like many ethologists of my generation, I was much influenced by Niko Tinbergen. Thanks to him, I found myself asking again and again, what is the function of this pattern of behaviour or process? It raised questions about development that I would have never considered otherwise. Why, for instance, does the sensitive period for sexual imprinting occur later in the life-cycle than the sensitive period for filial imprinting? With Nick Humphrey, I founded a project on sociobiology and behavioural ecology in my Cambridge College, King's. The project was highly successful and eventually led to a conference and then to a book. In my chapter in that book I started to explore the active role of behaviour in evolution (Bateson, 1982). This interest has grown over the years

(Bateson, 2006a) and has recently led to collaboration with two Russian colleagues, Kostya Anokhin and Misha Burtsev, developing models of how plasticity might have evolved and then drive evolution.

2. What does your work reveal about biological evolution (or evolutionary aspects of your field) that other academics, citizens, philosophers or biologists typically fail to appreciate?

Because of my own interests in development, I have felt for a long time that the gene-centred view of evolution was misleading. Furthermore, I felt that this view had an unfortunate effect on how the lay public viewed development. Metaphors such as "genetic blue print" acquired a wide currency and it became commonplace to refer to "genes for characters". In order to counter this trend, I co-authored a book with Paul Martin called "Design for a Life" which was intended for a general audience (Bateson & Martin 2000). The aim was show how an individual's development could be viewed as a dynamic process involving the interplay between the individual and its environment while at the same time being an adaptive product of Darwinian evolution.

3. What, if any, practical and/or social-political and/or moral obligations follow from your work on evolution?

I am not a great believer in human universals and take the view that even such widespread proscriptions as the incest taboo can be interpreted in a different way from that which is commonly accepted. I see these taboos as a product of a strong disinclination by most people to choose sexual partners who are familiar from early life coupled with the imposition of conformity on those who might behave differently (Bateson 1989).

The understanding of how early experience provides a weather forecast of the conditions into which an individual is likely to enter as an adult are particularly important for the design of public health programmes in rapidly developing countries in which the mismatch between an individual's phenotype and its environment is often large (Bateson 2001). The understanding of developmental processes is also central to educational programmes since habits of a life-time are formed early in development (Bateson & Martin 2000)

4. What do you see as the most interesting criticism against your position in the biological or philosophical discussion of evolution?

I have been critical of the thinking of my old friend Richard Dawkins when he conflates his teleological selfish gene approach with a mechanistic view of the gene as a programmer of development. In his turn he has criticised me for being obscurantist (Dawkins 2004). An old joke of mine about the bird being the nest's way of making another nest elicited an irritated denunciation: "... Bateson's nest joke and others of this kind are not funny. There may be backwards arrows in all sorts of other senses but, in the sense that specifically matters for Darwinian evolution, the causal arrow of biological development from genotype to phenotype really is a one-way arrow." Earlier on, Dawkins (1982) had accepted that we might have been at cross purposes and wrote: "As is so often the case, an apparent disagreement turns out to be due to mutual misunderstanding. I thought Bateson was denying proper respect to the Immortal Replicator. Bateson thought that I was denying proper respect to the Great Nexus of complex causal factors interacting in development." His ironic reference to the Great Nexus (not a phrase I have ever used) was not intended to be complimentary, however, and it was that phrase that led to his charge that I am an obscurantist. In my defence, I have always tried to generate explanations that will render complicated phenomena more easy to understand (Bateson 2006b). I would certainly agree with Dawkins that being complicated for its own sake has no merit. On the other hand, explanations are worthless if they do not bear some relation to real phenomena. Understanding how the parts relate to each other is a precondition to understanding process and understanding process is the precursor to uncovering principles. I accept, however, that some tension will inevitably exist between those who resist sweeping generalisations and those who emphasise the grand simplifying idea.

5. With respect to present and future inquiry, how can the most important problems concerning evolutionary theory (or evolutionary aspects of your field) be identified and explored?

First, I believe that Darwin's metaphor of "natural selection", which did so much to win over the nineteenth century intelligentsia to his evolutionary theory, has become an encumbrance in the 21st century because it implies that an external hand "chooses" between individuals. As we learn more about niche construction and the adaptability driver of evolution, we begin to appreciate how important are the activities of the organism in affecting the evolution of its descendants. Second, the new science of epigenetics will change the way that we think about the third main component of Darwinian theory, namely the mechanism of transmission of characteristics from one generation to the next. By providing a means of generating intra-familial heritability, it allows more time for favourable genetic mutation to occur. Third, I believe that Darwin's great theory was more to do with the origin of adaptation than it was to do with the origin of species. The idea that species are only formed by slow continuous change is looking increasingly dubious. Focus on hybridisation, freaks and the origins of chromosomal reorganisation will reveal how discontinuities between generations can arise.

References

Bateson, P. (1982a) Behavioural development and evolutionary processes. In King's College Sociobiology Group (Eds.) *Current Problems in Sociobiology*. Cambridge, Cambridge University Press.

Bateson, P. (1989) Does evolutionary biology contribute to ethics? *Biology and Philosophy*, **4**, 287-301.

Bateson, P. (2001) Fetal experience and good adult design *International Journal of Epidemiology*, **30**, 928-934.

Bateson, P. (2002) William Bateson: a biologist ahead of his time. *Journal of Genetics*, 81, 49-58.

Bateson, P. (2006a) The Adaptability Driver: links between behaviour and evolution. *Biological Theory: Integrating Development. Evolution and Cognition*, 1, 342-345.

Bateson, P. (2006b) The nest's tale: a reply to Richard Dawkins. *Biology and Philosophy*, **21**, 553-558.

Bateson, P. & Martin, P. (2000) *Design for a Life: How behaviour develops*, Viking paperbacks.

Dawkins, R.: 1982, *The Extended Phenotype*. Freeman, Oxford.

Dawkins, R.: 2004, Extended phenotype - but not too extended. A reply to Laland, Turner and Jablonka' *Biology and Philosophy* **19**, 377-396.

Irvine, W. (1955) *Apes, Angels, and Victorians*, London, Weidenfeld & Nicolson.

2
John Tyler Bonner

Emeritus Professor
Department of Ecology and Evolutionary Biology
Princeton University, USA

My first brush with evolution came when I was about 15. I had already made up my mind that I wanted to become a biologist, and without much guidance stumbled into one biology book after another. I decided to give *On the Origin of Species* a try which was a big mistake; I urge all 15 year olds to wait until they are at least 20. It is not a book for the callow. The experience did not discourage my biological interests; I just thought there were better things for me to read, and there were. In my junior year in college I finally fixed on developmental biology as being the subject that appealed to me most, and since I was also captivated by lower organisms, I combined the two and started to work on the development of cellular slime molds. This is what started me on a way that ultimately brought me to evolutionary biology.

These slime molds have a most peculiar life cycle. Unlike most organisms they eat first (as separate amoebae in the soil) and then develop into a multicellular organism. Most other living things eat and grow at the same time. As I continued—with delight—to experiment on their development to see if I could find out some of the mechanisms of how they develop, I was constantly reminded of the oddness of their life cycle. (When I started as a student it was striking that most normal biologists had never heard of anything so weird, but now it is part of every high school biology text.) And I was also fascinated by other lower organisms that had equally peculiar life cycles; I became more and more fixated on comparative life cycles. One day it dawned on me that while we tend to think of organisms as adults, or young, or even embryos, in fact they are life cycles. A dog is a dog from the fertilized egg to the decrepit adult; our habit of thinking of a dog as a slice in time is obviously convenient, but this short cut does not reflect the true nature of the dog. A dog is a life cycle, as indeed is a slime mold, although their cycles differ in significant ways.

This led me to a fascination with the matter of size in living organisms. The fertilized egg of the dog is minuscule compared to the adult beast, which means that there is a dramatic size change over time during the cycle. It makes no difference if we are considering an elephant or a humming bird; the principle is the same although the difference is great. It may seem an odd route, but this is what brought me to evolutionary biology.

By now I was a totally convinced Darwinian, and indeed had gone so far as to be able to absorb *On the Origin of Species* and appreciate the message. (I never feel as though I have fully encompassed his message, because every time I dip I find something new and often truly modern that anticipates present day thinking.) So I began to wonder why has natural selection produced life cycles and why has it produced organisms of different sizes.

This amounts to asking why has there been a selection for development. Darwin appreciated the need to consider the role of development in the *Origin*, as did many others that followed, the most conspicuous of which was Ernst Haeckel with his notorious "ontogeny recapitulates phylogeny." He has been attacked for the last 100 years because things are more complicated, and there are indeed many interesting matters among those complications, but I am fond of championing the underdog. While his so-called recapitulation theory is a blatant oversimplification, it nevertheless does underline a basic truth: the stages of development can reflect stages of evolution. In many ways this is the pre-molecular cornerstone of the modern movement of evo-devo.

At this point I went off on a tangent. I have always been fascinated by the parallels between development and animal societies. There are so many similarities that one readily gives insights into the other. This is a theme that continues to be very much on my mind. Both are integrated by a signaling system that coordinates them and makes them into cohesive wholes. However, there is one important difference: animal societies are capable of transmitting signals of a unique sort; they can pass on information from one individual to another that is non-genetic, that is, they are capable of culture. This had been pretty much in the domain of anthropologists who considered culture a strictly human occupation, but there was an increasing awareness that great apes and other animals also had a kind of primitive culture. For instance, many animals used tools for extracting food and this is not genetically determined, but passed on from one individual to another by learning, by imitation. For me this led to an exploration into the evolution of culture—what were the roots, the origins of culture?

But size became the increasingly important element in my mind. There are so many indications that natural selection for size plays a ruling role in evolution. Not only did life start off with minuscule prokaryotes, and as the geological eras succeeded one another bigger and bigger animals and plants appeared. Selection for size increase has been a major component in the history of life on earth. It led to multicellularity and it led to animal societies, as well as increasingly larger multicellular forms and increasingly larger societies.

This is an account of my trajectory into evolutionary biology. Let me now outline the progress of evolutionary biology itself for it has come a long way since Darwin. This is a familiar story so I shall be very brief. The discovery of Mendel's genetics and how it fused with natural selection in the form of population genetics was the great advance in the early part of the 20th century. The world of biology was turned upside down mid-century with the mega discovery of molecular genetics and this has had a reigning influence on evolutionary biology. This made it possible to revisit the whole matter of family trees that had been based on morphology, but molecular phylogeny could provide, and is providing, a far more accurate, and infinitely more detailed tree of life for all groups of living organisms. Even more exciting was the realization that molecular biology could reveal the steps of development, and again in rich detail, with the result that developmental biology has taken on a new and life. And the more we learn about development, the more we learn about evolution.

Besides those activities of genes in populations and in development there have been two other movements in recent years. One is the realization that in analyzing development one cannot ignore physical forces such as movement and adhesion; genes are just one of the essential players; it is not all in the genes. The other is that genes do their job in development in concert: one gene producing one protein is not enough; rather there is an array of genes, some that contribute to the steps of development by regulating the activities of others. There is a complex network that controls each step in development.

To return to evolution, this complexity has arisen by natural selection. Initially there must have been a simple first step, and as I have argued in the past, over millions of generations that step has been added to with alternate pathways, and with modulating genetic pathways that all lead to making each developmental step more certain to persist and more reliably controlled. The result is that the development of any organism living today has evolved from simplicity to complexity over endless stretches of time.

The new molecular evo-devo has brought us two new things, one of which is wonderfully illuminating, and the other rather muddy. The former is best illustrated by the HOX genes that lay down the body plan for animals. Here is a simple bit of genetic information that links all animals; it establishes a principle of development that is so fundamental that it rises above the complications. It provides a sort of top down view of the control of development, a dazzling and fascinating view.

The muddy side of modern evo-devo are all those complications. The more one analyzes the steps in the development of any organism, the more details—often interrelated—emerge. The result is that instead of simplifying a developmental process one complicates it. What can be done? One approach that holds much promise is systems biology which tries to find integrating principles that organize the details. It is the opposite from reductionism, and by means of mathematical models one can make order out of the chaos of the details. A comparison can be made to the field of ecology where there was a morass of detailed observations, and in the mid twentieth century, Robert MacArthur and others found simple mathematical models that shed dazzling light on that morass. No doubt that can be achieved for the evolution of development too. However, it is a paradox that what we are looking for are the initial bare bones—of how it all started before the details have evolved. I have tried this approach in the past, but the difficulty is that by necessity these are highly speculative thought experiments. No one was there to observe it when it all started millions of years ago. Although I sometimes wonder if it is not starting again today in some contemporary reinvention of multicellularity, but we have no way of finding it.

There has been a flurry of recent commentary that the new synthesis of the 1930's needs to be superceded and we need a new, new synthesis. My feeling is that we have it. Those old ideas of the past were tremendously important and were the foundation of all that has happened since then. What we have added is the appreciation that genes are not everything (although they are central and incredibly important), but physical forces are equally so. Genes do not exist in a vacuum; they impinge of material substances, and those substances set the rules of what can and what cannot be done. And the genes are further restricted by what was built in the past. Each new species does not start with a blank slate, but is encumbered with all the constructions that went before. And there are many more important advances since the 1930's. Our insights will keep progressing: the evolutionary new synthesis will continue to become newer into the future.

3
Terrence W. Deacon

Professor
Department of Anthropology (Biological Anthropoogy Division), and
The Helen Wills Neuroscience Institute,
University of California, Berkeley, USA

---

1. Why were you initially drawn to discussions and research on evolution (or evolutionary aspects of your field)?

At a very early age, children' books depicting the stages of biological evolution, and of course dinosaurs, captivated my attention. Although I had heard the Adam and Eve story, even at a young age I could distinguish a story from an effort at explanation. And the explanation was so much richer and more fascinating. As a child I was equally interested in the possibility of space flight and the nature of the cosmos, and the power of science to serve as the handmaid of curiosity. My interest in both biology and the physical sciences has always been a powerful influence and in my early college years these interests began to merge as I found many writers exploring the problem of the origins of form—including the spontaneous appearance of inorganic forms such as snow crystal and basalt columns, and of course biological forms. In the mid 1970s I came under the spell of a number of thinkers struggling with this issue in its many diverse manifestations, such as Anderson, Ashby, Bateson, Bertallanfy, Bohm, Koestler, Pattee, Polanyi, Shannon, Simon, Waddington, Weiss, Wiener, and others. In the late 1970s I also became aware of the work of the American philosopher Charles Sanders Peirce, and especially the way his work cut through the confusions of mind/body dichotomies. It was at that point that I began to see that there must be a relationship between information in biology and in the cognitive sciences.

This intuition, that studying the one can inform the other, has followed me to the present and motivates me still. Thus, my post graduate studies continually swung back and forth between cognitive development, comparative neuroscience, and evolutionary biology, and ended up leading me to a PhD thesis in 1984 based on a neural tracer study in monkeys of the homologues to language relevant brain structures. A decade later amidst studies of the development of neural circuits, using cross-species fetal neural transplantation, I attempted to bring my divergent interests and studies together in my book titled "The Symbolic Species: The Coevelution of Language and the Brain". In it I attempted to show how bringing together insights from semiotic theory, neural development, and multilevel evolutionary theory could illuminate the mystery of the origins of language. Though more of a model hypothesis than a review, I believe it still represents the prototype of the multidisciplinary form of argument that must be brought to bear on this topic. This work also drove home to me the limits of both simple genotype-phenotype assumptions and simple neoDarwinian accounts of function for such higher-order adaptations. It forced me to confront the problem often described as "emergence', and I have spent much of my subsequent work struggling to develop a general scientific theory of such changes in evolution.

2. What does your work reveal about biological evolution (or evolutionary aspects of your field) that other academics, citizens, philosophers or biologists typically fail to appreciate?

a. That evolution is due to the interdependence of both selection and self-organization processes, and that these play quite distinctive and complementary roles in both development and phylogenetic evolution.

b. That relaxation of stabilizing selection (e.g. as might be due to invasion of an unoccupied niche or domestication) tends to release constraints on self-organizing mechanisms (a) involved in development allowing morphogenetic interactions that are otherwise prevented, and which can produce traits that can come under selection for the novel higher-order synergistic consequences

c. That this (b) is the principal source of higher-order synergies and thus complexity in evolution.

d. That this synthesis is key to developing a scientific theory of emergence, that is not merely a critique of eliminative reductionism, but rather a theory of the common processes underlying the generation of the causal complexity of the biological and mental worlds.

e. That symbolic communication (e.g. protolinguistic forms of communication) maintained by social transmission, provided a source of selective influence on the evolution of human neurological organization and cognitive capacities, and that this has been a principal factor in the divergence of human cognitive and emotional systems from those of other species.

f. That integrating the insights gained from the study of self-organizing processes with the insights gained from the exploration of the link between selection and development, will lead to reformulation of the notion of biological information, re-grounding it in more basic dynamical principles. It is currently one of the dominant assumptions in evolutionary biology that molecular template replication is sufficient to explain evolution and is what warrants claiming that nucleic acid molecules convey information, but this has systematically ignored thermodynamic and morphogenetic presuppositions of the concepts of replication and reproduction. When these are fully taken into account, neither molecular replication nor molecular information can be seen as primary attributes of life or evolution. Inevitably, the information-bearing properties of DNA and RNA will be understood as evolved specializations, and thus derived from antecedent dynamical processes capable of evolving and reproducing irrespective of these functions.

That the eventual understanding of how information-relationships emerged from self-organizing and self-reproducing dynamics that lack specific replicated information-bearing components—and thus that evolution by a form of natural selection does not depend on this property—will not only advance the study of the evolutionary process and research into the origins of life, but it will also contribute to a general theory of information.

3. What, if any, practical and/or social-political and/or moral obligations follow from your work on evolution?

I think that both practical and even moral consequences follow. Understanding the hierarchic dependency of self-organizing processes on non-equilibrium thermodynamic processes (and constantly perturbed or otherwise nonlinear computing processes) and of adaptive processes on self-organizing processes, will eventually provide tools for understanding how to better interact with complex adaptive processes in many practical domains, from medicine to economics. It will also someday underpin the development of true thinking devices, with real semiotic functions, and conceptions of the world they work in. This of course has enormous moral implications. But more importantly, this is the path to expanding the natural sciences to be consistent with an understanding of meaning and value as real efficacious physical relationships. Of course such a synthesis is a long way off, though it is in my opinion the only way toward a re.

I am particularly disturbed by the anti-science propaganda war that has been waged against evolutionary research by fundamentalist religious organizations, both Christian and Muslim, and how this has exemplified a growing anti-intellectualism anti-science sentiment growing around the world. This is a threat to civilization, not just freedom of thought. Though this is often portrayed as a science-versus-religion battle over an explanation of the way things are, I think that the stakes are somewhat more basic. It is a battle over the ontological status of teleological processes and relationships—e.g. meaning, purpose, consciousness, value, ultimacy—where evolutionary theory is seen as the exemplar of a form of explanation that completely denies teleology and ultimate meaning. It is, in this sense, viewed as the enemy of value, the promoter of moral relativism, and the foundation of crass secularism, that religious communities find most threatening.

This is further abetted by radical postmodernist views currently popular in the social sciences and humanities that promote the belief that scientific knowledge is no better grounded than any other claim to opinion. This too is a reflection of growing anti-establishment sentiments that have come to mistrust the scientific community as it has become increasingly influenced by financial interests at the expense of open inquiry. This is unquestionably a source of bias influencing the sciences, and it is indeed a cause of great concern that the increasing power of government and industry influences dictate the reporting of "acceptable" findings of scientific research.. Curiously, evolutionary biology remains a field that has only minimally been able to gain the attention of major political and economic interests, is largely thought to be of little practical economic use, and yet is a major target of criticism and politicized attacks. This is in large part due to the perception that biological explanations are often seen as challenging social constructivist views which are ideologically important to the far left, and the perception that religious fundamentalists can be manipulated by attacks on evolutionary theory to aid the consolidation of political power by the far right. In this way, evolutionary biology is a useful target for both extremes.

These engines of misinformation have unfortunately also generated a tendency toward reactionary defence within the biological community. And this has contributed to a growing unreflective closed-mindedness within the scientific community as well. Thus even theoretical speculations that only modestly diverge from standard modern neoDarwinism have a hard time gaining mainstream attention because it is feared that these provide "support for the enemies of science." Luckily, despite the attraction of politicized extremes, simplistic accounts that play well in the popular media, and being out of the mainstream flow of "big money", the field is currently experiencing unprecedented efforts to expand the current dominant paradigm, and develop a more sophisticated theory of evolution.

4. What do you see as the most interesting criticism against your position in the biological or philosophical discussion of evolution?

I can't say that I have found any criticisms to be of particular interest, though there are many that offer serious challenges and attractive alternatives. Most intriguing (perhaps because of their opaqueness) are theories that rely on quantum-mechanical effects to explain especially information–related phenomena. The many alternative scenarios proposed to deal with the origins of life and of the evolutionary process most catch my attention currently, because they are forced to provide an explanation of evolution, and not merely assume it. Exploring this problem cannot but help our understanding of the evolutionary process at every level.

5. With respect to present and future inquiry, how can the most important problems concerning evolutionary theory (or evolutionary aspects of your field) be identified and explored?

The increased interest in developmental processes and the ways that biological form and function are generated will be one of the most important contributors to evolutionary biology in the coming decades. Also the increased use of sophisticated computer modelling and agent-based simulations to test the consistency of the evolutionary logic of various theories will be critical. In addition, simulation and laboratory approaches to the problem of the transition from non-life to life will become increasingly important sources of insight for evolutionary biology in general, and will show themselves to even be important for the development of cognitive neuroscience. Finally, there remains considerable work to be done by philosophers of science to broaden our conceptual understanding of emergent phenomena, informed by these advances in evolutionary biology and the physics of self-organized phenomena. This includes a fundamental rethinking of concepts of information and meaning.

4
Daniel C. Dennett

University Professor, Co-Director, Center for Cognitive Studies

Tufts University, USA

---

1. Why were you initially drawn to discussions and research on evolution (or evolutionary aspects of your field)?

When I was a graduate student in philosophy at Oxford in the mid sixties, I knew very little about evolution, beyond the general shape of the theory and its bottom-up way of getting design without a designer, but when I confronted the question of how minds (or brains) could possibly (non-mysteriously) do the work they did, it seemed to me to be quite obvious that they *had* to work by evolutionary principles. Darwin had provided a model of *design-improvement* without need of an intelligent designer, and what was a learning brain but a material thing that could somehow, in virtue of its design, *redesign itself* in beneficial ways? Learning has to be a kind of self-design, and evolution by natural selection is not just our only model of such a process; it shows quite clearly why there couldn't be an alternative.

Neural nets, as many have seen, are composed of elements that fairly beg to be treated as the loci of selective pruning and growth, all accomplished by uncomprehending ("mechanical") processes. What could, in principle, emerge from these processes would be networks with ever more refined and elaborated cognitive competences, reaching out into the world via sensors and effectors that would enable the controlling states to come, in the fullness of time, to be *about* things and events and properties in the world.

I developed this thesis in my Oxford DPhil dissertation, and was heartened to learn that one of my examiners, the British neuroanatomist J. Z. Young, was working on a similar theory of learning. My version was published in my first book, *Content and Conciousness*, in 1969, and I soon learned that there were other versions around, from Thorndike and Skinner, and from Darwin himself, among others. It was the behaviorist versions that particularly drew my attention, and in 1974, I published "Why the Law of Effect Will Not Go Away," which argued that while the behaviorists had been right to see learning as an evolutionary process, they had missed the fact that there could actually be a cascade of ever more sophisticated intra-cranial evolutionary processes. Not just Skinnerian creatures, who were blessed (by evolution) with a system of reinforcement that enabled them to engage in trial-and-error in the environment, which would improve the adaptiveness of their responses during their lifetimes, but "Popperian" creatures who could engage in *off-line* trial-and-error in their heads, using an inner environment of selection to bias their first choices for real world action in adaptive directions. (I cited Popper in the 1974 article, but I didn't adopt the term "Popperian creature" and the even more sophisticated "Gregorian creature" until *Darwin's Dangerous Idea* in 1995.)

So the idea of learning as a variety of evolution in the brain was my entering wedge, and at the outset I barely understood the theory, and had read very little beyond high school textbook fare and some of Stephen Jay Gould's collections of essays. (*Ever Since Darwin* is still my favorite, but Gould himself renounced some of his best essays in that volume.) But the more I learned about evolution, the more implications and applications in philosophy of mind I saw. Dawkins' *Selfish Gene* was a major influence, and soon it led me to George Williams, Bill Hamilton, Robert Trivers, John Maynard Smith, and many others. I saw more in Dawkins' concept of memes than most other readers, and decided to push further in the directions he had pioneered, and this has occupied an ever larger place in my perspective on minds. As I sallied forth with my Darwinian formulations, I became acutely aware of the many anxieties and visceral rejections of Darwinian thinking in my colleagues, not just in the humanities but in the social sciences, and this phenomenon, I decided, deserved focussed treatment (in two senses—both diagnosis and therapy!) The result was *Darwin's Dangerous Idea*, in which I attempted to lay bare both the power of Darwinian thinking across the spectrum of all human thought, and the sources of misgivings. This has proven to be, I think, my most valuable contribution to thinking about evolution so far.

2. What does your work reveal about biological evolution (or evolutionary aspects of your field) that other academics, citizens, philosophers or biologists typically fail to appreciate?

It seems to me that there is still massive inertial resistance to evolutionary thinking in the humanities and, to a lesser extent, in the social sciences. The very idea that artistic or scientific creativity, for instance, or ethical thinking could be accomplished by "mechanical" networks trained by intra-cerebral processes of trial-and-error strikes too many thinkers as, in a word, obscene. They are driven by that old Cartesian itch, the hankering for a *res cogitans* that is not beholden at all to the material world of causation, and even as they abandon frank dualism as an incoherent and deeply retrograde option, they continue to avert their eyes from what strikes me as just obvious: minds are nothing more than what brains do, and they do what they do by non-mysterious means, which can only be explained by their *design*. Brains have no wonder tissue, no magic ingredients that stomachs and kidneys—and leaves and tree trunks—don't have; their working parts are (mainly) proteins and other polymers that play deliciously sophisticated roles in larger systems. All this structuring had to be *composed* and *refined*. These nano-robots and the micro-robots and macro-robots they compose all had to be engineered—by natural selection. It isn't a question of evolutionary thinking usurping the topics and explanations of the humanities and social sciences; it is a matter of evolutionary theory *grounding* the levels of explanation employed by these disciplines. What many in the humanities seem not to realize is that they *need* an evolutionary account to underpin their perspectives. The idea that the *Geisteswissenschaften* and *Naturwissenschaften* can co-exist without any uniting theory is as forlorn as the old idea that sublunary and superlunary physics were entirely distinct. (Medieval thinkers took it on Aristotle's authority that the physics of things beneath the moon (on earth) was entirely different from the (in principle unknowable) physics of superlunary regions.)

3. What, if any, practical and/or social-political and/or moral obligations follow from your work on evolution?

The hidden agenda for much of the anxiety about evolution stems from the dimly conceived notion that if we turn out to be merely material organisms with merely evolved brains that make decisions with the aid of nothing but neural mechanisms designed by natural selection, we cannot have free will, or moral responsibility, or moral worth, and life would have no meaning. This is, I have argued, entirely mistaken. Not only is evolution not the enemy of free will; it is its salvation. The only way we can understand how any variety of free will can exist is to see it as a recently evolved characteristic of only one species—*Homo sapiens*—an elaboration of the simpler forms of freedom enjoyed by the bird and the dolphin and the ape. My 2003 book, *Freedom Evolves*, attempts to set out in great detail the arguments for this conclusion. Moreover, it is emerging that evolutionary theory casts very specific light on such otherwise baffling phenomena as altruism, ethical intuition, loyalty, love, honor. It does not "explain away" these cherished phenomena; it explains how they can exist in a world of atoms and energy. The price is not onerous: one must abandon a few traditional but ill-considered intuitions about these phenomena .

4. What do you see as the most interesting criticism against your position in the biological or philosophical discussion of evolution?

One of the most interesting criticisms I have encountered recently has been developed by Peter Godfrey Smith at Harvard. I have been arguing for years that we should not just acquiesce in, but positively exploit, talking about the *design* in nature, and in particular the *free-floating rationales* of those designs. Thus the engineers' rationales for the shape of an airplane wing are, in the main, articulated, represented, discussed, criticized. Engineers are reason-representers. Mother Nature's rationales for the shape of birds' wings are very much the same—but not *represented* anywhere. There are *reasons*, I want to say, for many if not all the features of organisms. These are perfectly good reasons that just don't have to have been represented until we reverse-engineers came along and figured them out! Godfrey Smith thinks this is extreme, gratuitous, dangerous, but he has not yet convinced me of this at all. Still, I have learned a lot from him about evolution, so I don't dismiss his objections—yet. This is work in progress.

5. With respect to present and future inquiry, how can the most important problems concerning evolutionary theory (or evolutionary aspects of your field) be identified and explored?

I hope—and expect—that the very important problems concerning cultural evolution are on the brink of being so clearly articulated that they can be jettisoned from the arena of "just" philosophy and become a solid topic of scientific research. After several decades of largely misbegotten antagonism and overstatement, a sane, useful and sound concept of memes will be seen as indeed comparable to the concept of genes (which, one must remember, has had its share of controversy even among evolutionary biologists and geneticists), and it will be included, *along with other related concepts*, in a rich and generative body of theory about cultural evolution. The work on evolution of languages, and on co-evolution (gene-culture interactions) will be consolidated, thanks to new quantitative methods, and the internet will provide the "model animals" for this science. For the first time in history, huge and important parts of culture reside in a format that is unproblematically measurable, recordable, searchable on a very large scale. The internet will prove to be not just the *Drosophila* or *C. elegans* for the science of cultural evolution; it will be like that dream of the neuroscientists: a living brain chronically tapped by billions of microelectrodes. We will drown in the data unless we develop good theories, but there are no shortage of candidates ready to be articulated to the point of testability.

5
Douglas J. Futuyma

Department of Ecology and Evolution
Stony Brook University, USA

Evolutionary Biology and its Social Role: A Personal View

1. How I Became an Evolutionary Biologist

In retrospect, it would appear that my path toward a career in biology started at the age of eight or earlier, on excursions with my father to the American Museum of Natural History and the parks of New York City, especially the New York Zoological Garden (the Bronx Zoo). Within a few years, I began to search the parks for insects to feed the lizards and frogs I kept at home, and so became aware of insect diversity. I spent countless hours at the zoo, learning about the diversity and distribution of terrestrial vertebrates, and finding amphibians and reptiles especially alluring. At the age of 13, I became an obsessive birder, and grew to know the habits, habitat associations, and migration schedules of the 250 or so species that can be readily seen in the city and its environs in the course of a year. In short, I became fascinated by the diversity of animals (and later, of plants) and threw myself into learning their taxonomy. I vividly recall reading and virtually memorizing one of the first scientific papers I encountered, Ernst Mayr and Dean Amadon's "A classification of Recent birds" (1951), which one of my birding mentors gave me.

I went to Cornell University with the intention of becoming a herpetologist, and reveled in the variety of traditional courses on vertebrates, insects, and plants – the so-called "ologies" that today are often scorned as though inferior, although I feel strongly that such training is indispensable for evolutionary and ecological biologists. I was dimly becoming aware that the essential framework for understanding biological diversity is evolution. Two of my teachers brought this awareness into focus: Charles G. Sibley and William L. Brown, Jr. Sibley described his research on speciation and hybridization in birds, and how he was embarking on a research program in which molecular similarities – electrophoretic profiles of egg-white proteins – would, he claimed, reveal phylogenetic relationships among birds. Brown, a charismatic figure who was the world's leading ant taxonomist, held forth, at a brown-bag lunch series and in his insect taxonomy course, on his ideas about speciation, which were at variance with Mayr's. I hardly knew what Mayr's position was, but I learned from Brown that one might challenge authority, and also that in order to speak knowledgeably about evolutionary theories, one should read a conceptually and taxonomically diverse literature. As I began to understand and to read about evolution, I realized that I must learn about genetics, and that I should take the course in population genetics offered by one of Dobzhansky's former students, Bruce Wallace. Although I found much of the material mystifying, I was convinced that this field was critically important to my understanding.

I enrolled in the Ph. D. program at the University of Michigan with the intention of pursuing my interest in herpetology, but I was becoming ever more intrigued with conceptual issues and questions. The greatest intellectual ferment in the zoology department at that time was among the ecologists, who occasionally touched on evolutionary questions as well. I made the difficult decision to abandon herpetology and became a student of the most self-consciously intellectual member of the faculty, Lawrence Slobodkin. Under his influence and that of Richard Lewontin, with whom I spent a summer of directed study at the University of Chicago, I decided to devise a research project that would combine ecology and evolutionary genetics. I decided on using experimental evolution to explore adaptation to interspecific competition in laboratory populations of *Drosophila melanogaster* and *D. simulans*, in an experimental design that also incorporated an effort to assess whether or not adaptation might be affected by founder effects (Futuyma 1970). This reflected my continuing interest in speciation as the source of diversity, and the great influence Mayr's *Animal Species and Evolution* (1963) had on me. This dissertation topic initiated my major research theme in later years, the evolution of interactions among species and its ecological consequences. It also illustrates a signal feature of my career, my interest in synthesizing diverse approaches and fields of study. When I began to seek a research project in the 1960's, a developing field of "population biology," intended to be a synthesis of evolution and ecology, was only in its first stages (e.g., Levins 1968, Lewontin 1968, MacArthur 1961), and there were rather few models of synthetic research programs. Both my dissertation and subsequent research were attempts at such synthesis.

2. My work and its reception

My contributions to evolutionary biology have been of two kinds. My primary research has been on the evolution of interspecific interactions, specifically on interactions between herbivorous insects and their host plants (including "coevolution" *sensu latissimo*). My other contributions have been in education, synthesis, and promotion of evolutionary science.

I think I was among the first empirical researchers to bring a population genetic approach to insect/plant interactions, and perhaps to interspecific interactions in general. My laboratory first looked for host-associated genetic substructure of insect populations (e.g., Mitter and Futuyma 1979), which led into studies of local adaptation to host plants and to the reasons for the evolution of host-plant specialization in insects (e.g., Futuyma and Philippi 1987, Futuyma and Wasserman 1981, Wasserman and Futuyma 1981). This work led me to develop a skeptical analysis of the hypothesis of sympatric host-associated speciation (Futuyma and Mayer 1980), and some of my students explicitly studied speciation (Funk 1998, Knowles 2000). Several students (F. Gould, J. D. Hare, A. J. Gassmann) for whom I was primary mentor or dissertation committee member carried over the "ecological genetic" approach into the applied sphere, viz. the management of crop pests. My laboratory also provided evidence that the evolutionary history of divergence in host affiliation of closely related species may have been guided in part by limitations on genetic variation in features necessary for adaptation to different plant species (e.g., Futuyma et al. 1995). This research was an explicit attempt to relate "microevolutionary" process to "macroevolutionary" pattern (Futuyma 1988). I also published a speculative suggestion (Futuyma 1987) that speciation may affect the long-term pattern of character evolution in such a way as to reconcile the pattern of punctuated equilibria (Eldredge and Gould 1972) with orthodox population genetic theory, when these were generally viewed as antithetical.

As an assistant professor, I was provided the opportunity to write a textbook on evolution. The result (Futuyma 1979) was well received, and led to other opportunities to indulge my interest in synthesis, such as becoming editor of *Evolution* (1981-1983). At the same time, during the Reagan presidency, religious conservatism began to reassert itself in the United States, and creationism in particular grew in strength. My contribution, *Science on Trial: The Case for Evolution* (1982), was one of the first attempts by evolutionary biologists to combat this threat to education and science. (I don't think I imagined at that time just how powerful and threatening the creationist movement would become, nor that many more biologists would realize the need to devote effort to the conflict.) As a result of these several activities, I have become increasingly engaged in syntheses and in helping to represent and promote evolutionary biology, both within and outside academia.

On the whole, I feel that my contributions in both of these roles have been recognized. Like most researchers, I have published work that I wish had received more notice. For instance, my hypothesis for reconciling punctuated equilibria and population genetics has drawn rather little attention among evolutionary geneticists, although paleontologists have cited it approvingly (e.g., Gould 2002). But since I haven't put very much effort into promulgating the idea, much less produced any supporting data, perhaps I should expect the tepid reception (assuming that the hypothesis has any value at all). My proposition, based on empirical evidence (Futuyma et al. 1995), that genetic variation may be more limited than is generally supposed and might therefore constrain adaptation, was viewed skeptically at first, but this view has recently been gaining more evidence and adherents (e.g., Blows and Hoffmann 2005). Sympatric speciation, on which I have taken a quite conservative stand (Futuyma and Mayer 1980, Futuyma 2008), continues to be a highly controversial topic, and while the position that I and many others have taken has been widely recognized, there are still many enthusiastic advocates of sympatric speciation. However, I feel that most of my work on interspecific interactions, both empirical and synthetic, has been accepted and appreciated. Certainly, my broader educational and advocacy contributions have been well received.

3. Practical and social implications

I have not devoted any personal research effort to the applications of evolutionary research on insect/plant interactions, but this is an extremely active field, in which both my own students and others I have influenced have been active. Insect pests have been shown to evolve resistance to many control measures, including both chemical insecticides and the use of (initially) resistant genetic strains of crops. Moreover, herbivorous insects are often used to control weeds, usually by importing an insect from its native land to control a weed that originated in that region. A critical issue is that such insects should be host-specific, and not damage other crops or native plants. Thus, evaluation of an insect species as a potential control agent should take into account not only its current degree of host specificity, but also the likelihood that it may adapt to nontarget plant species (Futuyma 2000). Many researchers who evaluate measures to manage pest insects and weeds now use the principles and methods of evolutionary genetics, as well as ecological principles, and have appreciably changed applied entomology within the last 30 years. Even more fundamentally, work on both pests and control agents requires that species be properly distinguished and characterized, an area in which both evolutionary genetics and phylogenetic analysis play an important role. I am gratified that one of my former students, Sonja Scheffer, now a researcher in the U. S. Department of Agriculture's Systematic Entomology Laboratory, has become the American authority on the systematics and evolution of a major family of flies that includes many crop pests.

My work on behalf of the field of evolutionary biology has much broader social implications. Evolution, together with the principle that all biological processes fundamentally consist only of chemistry and physics (to the exclusion of any immaterial "life force"), is one of the two unifying principles of biology. Every aspect of biology is informed by posing it in an evolutionary perspective, and of course there is a long history of seeking implications of evolution in fields such as anthropology, psychology, and philosophy. The utility of evolutionary theory is dramatically illustrated by the increasingly important role it plays in developmental biology, molecular biology, and genomics. Evolutionary theory has many applications in medicine, agriculture, fisheries, conservation, and many other fields, and evolutionary biologists are becoming increasingly conscious of their social role, as reflected in the recent establishment of the journal *Evolutionary Applications*. I have played a small but satisfying role in encouraging this awareness (Futuyma et al. 1998), both within and outside professional evolutionary circles.

Most importantly, I believe, evolution is the locus of confrontation between two world views: science, rationalism, and an insistence on evidence versus a rejection of science in favor of unquestioning faith. In my view, there is no necessary conflict between accepting evolution and holding some forms of religious belief. (Of course, there is irreconcilable conflict between science and some specific religious beliefs, such as special creation of species in their present form.) Nevertheless, more than half of the American public does not accept evolution, and skepticism about science extends (even within the Bush administration, now thankfully at an end) to other issues as well, such as the anthropogenic causes of global climate change. Dismissal of science is more extreme in some other countries where religion is an even more dominant force than in the United States, and has been blamed for slow economic and social development in, for example, some Islamic countries. Adoption of an evidence-based, rational world view is critically important for technological and economic progress and, even more importantly, for the rejection of authoritarianism on which intellectual and even personal freedom depends.

To maintain a creationist stance, whether in the Biblical literalist form that prevailed in the 1980's or in its current guise of "intelligent design," is to reject evidence-based decision-making and rationalism. In *Science on Trial*, I wrote at some length about the nature of science, why creationism is not a scientific hypothesis, and how it threatens science as a whole. Subsequently, indeed, the "wedge document" was uncovered, which showed that the intent of "intelligent design" advocates was not only to displace evolution from science curricula, but to recast the sciences generally in a theological framework (Forrest and Gross 2004). The conflict between evolutionary science and creationism, then, is the front line at which science as a whole needs defense. This has become widely recognized, and is stressed in such publications as the U. S. National Academy of Science's publication *Science, Evolution, and Creationism* (2008) and in its designation of 2009 as the Year of Science (originally the "Year of Public Understanding of Science").

4. Differences of opinion

For the most part, I have not taken stands on evolutionary topics that differ strongly from prevailing views. However, I will cite two issues on which I have had disagreement with some of my colleagues in evolutionary biology.

One of the more conspicuous and sometimes strident debates has concerned the extent to which human cognitive abilities and behavior are genetically influenced and are a product of our evolutionary past. The debate was epitomized by the conflict between E. O. Wilson, who introduced the term "sociobiology," and his Harvard colleagues Richard Lewontin and Stephen Jay Gould, who opposed what they considered genetic determinism and went on to criticize the "adaptationist program" on which they thought it was based (Gould and Lewontin 1979). For a long time, my sympathies, which found some expression in my textbook, were mostly with Gould and Lewontin, partly because I was convinced by their charges that much sociobiological theorizing was unrigorous and that the data on heritability of characteristics such as IQ score were weak. I now recognize that my political sympathies almost surely colored my reactions as well. I now feel that with the passage of time, and perhaps in part because of the criticisms leveled by Gould, Lewontin, and others, studies of heritability have become quite rigorous, and that "behavioral ecology" (incorporating and improving on some sociobiological hypotheses) has proven its worth in studies of nonhuman species, has adopted higher standards of rigor, and can legitimately be extended to humans. A critical point is that as a general mode of scientific procedure, we are inclined to accept hypotheses insofar as they make successful predictions of new data. I now think that behavioral ecology has met this standard, and I accept that evolutionary psychology and its relatives may be able to do the same. Moreover, I now think that liberal values need not be threatened by evidence that human behaviors have an evolved, genetic basis, and that whether this be an appealing conclusion or not, scientific evidence, obtained as objectively as possible, should be gathered and faced openly.

A larger issue, on which I disagree with a minority of evolutionary biologists, is the extent to which current evolutionary theory is correct and sufficient. In this context, I have unquestionably been conservative. Many scientists aspire to intellectual revolution, to engineer a "paradigm shift" according to Kuhn (1962). My position has consistently been that the Synthetic Theory of evolution, as developed during the Evolutionary Synthesis (Mayr and Provine 1980) of the 1930's and 1940's, is solid, and needs not replacement but extension. The closest approximation I have seen to a paradigm shift in evolutionary biology is the neutral theory of molecular evolution, as developed by Kimura (1983) and developed further by others. But even this did not overthrow the claims of the Synthesis: it used and developed Wright and Fisher's theory of genetic drift, and applied this principle to a new kind of data, molecular variation, while accepting that phenotypes probably evolve primarily by natural selection.

Other attempts to "revolutionize" evolutionary theory have been largely unsuccessful. After Williams, Maynard Smith, Hamilton and others established that few, if any, adaptations require explanation by selection above the level of genes and individual organisms, D. S. Wilson (1983; Wilson and Wilson 2007) fought to revive a form of group selection, but I agree with many workers that his hypothesis is an alternative description of kin selection. Theory based on selection at the level of individuals and genes (including kin selection) has been a highly effective basis for explaining animal behavior, life histories, genetic systems, and much more. Similarly, some paleobiologists, S. J. Gould foremost among them, urged a major revision of evolutionary theory by proposing that populations generally cannot undergo adaptive evolution except during speciation (punctuated equilibria), that morphological evolution may occur by major steps based on the amplification of slight genetic changes by developmental processes, and that the history of life has been shaped largely by the differential survival and replication of species and higher-level clades. The result was a salutary recognition of the importance of development for understanding the evolution of form and of differential speciation and extinction rates in determining the distribution of phenotypic diversity, but these changes amounted to a revived recognition of old themes that had been neglected after evolutionary genetics took center stage in the Evolutionary Synthesis. The pattern that Eldredge and Gould (1972) called punctuated equilibria might be real, and certainly these authors served the field well by pointing to stasis as a phenomenon that requires explanation, but their proposition that populations cannot evolve except when they speciate, and that anagenetic change must therefore be produced by species selection rather than natural selection within populations, has been thoroughly rejected, and Gould himself abandoned it in his final work (Gould 2002).

As an author of a textbook, I have perhaps been pushed into a conservative position, because it does students no good service to promulgate paradigm shifts that then must be renounced or quietly omitted from a subsequent edition. But especially in view of such histories as I have just recounted, I also feel that we should not reject a position, at which our predecessors and colleagues have arrived by hard work, without skeptically scrutinizing possible replacements. Thus, I have, so far, given little credence to, and in some cases have expressed skepticism about, a few challenges to "received wisdom." I have already alluded to my continuing skepticism that species are commonly formed in sympatry. I have been critical of the proposition that nongenetic forms of inheritance (e.g., DNA methylation) have played a significant role in evolution, and will be until this rather plausible hypothesis is backed up by evidence. I see no need to replace the Synthetic Theory, which rates selection and drift of genetic variants as the fundamental, primary process of evolution, by proposals that the intrinsic dynamics of development sufficiently explain biological form, as some developmental biologists proposed (Futuyma 1984). I am skeptical of my friend Mary Jane West-Eberhard's (2003) proposal (reviving that of some older authors) that random phenotypic variants, evoked by environmental conditions, pave the way toward adaptation, and only later become "genetically accommodated" by a process like Waddington's genetic assimilation. Such a reversal of the accepted evolutionary process, which has been so fully documented, requires strong evidence, and I am not aware of any to date. (See de Jong 2005). I am enthusiastic about evolutionary developmental biology, which is providing a long hoped-for mechanistic basis for understanding the origin of the variation and the nature of constraints that help to shape phenotypic evolution, but at least so far, I see it as a necessary supplement to a correct but limited theory of genetic change.

Charting the future of evolutionary biology.

Evolutionary biology needs no suggestions about where to look for important, challenging problems. The field is in a state of excited ferment and progress on many fronts, including most conspicuously evolutionary developmental biology, evolutionary genomics, and phylogenetic analysis of evolutionary history, but also speciation, evolutionary ecology, evolution of behavior, and many other topics. New problems will continue to arise, and old unanswered questions will be revived, on the basis of new information, technology, theory, and synthesis:

Information. Much of the astonishing progress that has been made in many areas of evolutionary biology is a direct consequence of the growth of "mechanistic," or "reductionist," areas of biology, especially molecular biology and, more recently, developmental biology. The deeper understanding of how organisms work provides a basis for more deeply understanding their evolution (so, for instance, we may ascribe a phenotypic difference among species not just to anonymous genes, but to particular genes with well understood modes of action). Moreover, it provides whole new phenomena to study from an evolutionary point of view. We could not study exon shuffling or noncoding DNA until we knew that they existed. As new phenomena (such as microRNAs) are revealed by mechanistic biology, they will add to the problems studied in evolutionary biology, which takes all biological phenomena as its province. I do not wish to imply, of course, that the only new information that will prove useful to progress in evolutionary biology is molecular in nature; subjects such as evolutionary ecology and behavioral ecology progress in part by synthesizing and adding to comparative data on the distributions and ecological characteristics of organisms, and phylogenetic inferences about the evolution of characteristics are strengthened by comprehensive data on the features of diverse species.

Technology. In the same vein, much of evolutionary biology now uses the technology that enabled the growth of molecular and other mechanistic fields of biology. The other major technology that has revolutionized our field is information technology. Two decades ago, it was hard to imagine that phylogenetic analyses might be performed with thousands of base pairs for hundreds of species. Presumably, technology twenty years from now will provide data and analyses that only a few people today can envision.

Theory. Entire fields of evolutionary inquiry have arisen and prospered in the last 40 years not because of technology or molecular biology, but because of the development of new theory. Conspicuous among these are behavioral ecology, life history theory, phylogenetic approaches to community ecology, and the theory of sex and recombination and related phenomena. Mathematical theory and conceptual development have also enabled progress on topics that involve molecular data, such as the development of coalescent theory and its role in interpreting DNA sequence data. Theory has often paved the way to empirical study in evolutionary biology, and will continue to be indispensable.

Synthesis. Some of the most exciting and arguably most important progress has arisen from the conjunction and synthesis of fields. Dialogue among researchers in mathematical genetics, empirical population genetics, systematics, morphology, and paleontology produced the Evolutionary Synthesis that established the framework of modern evolutionary biology. The ongoing union of ecology and population genetics gave rise to evolutionary ecology, in which such diverse topics as life histories, niche breadth, and interspecific interactions are explored. Molecular evolution has joined with phylogenetics and provides insights into diverse problems, from dating historical events to natural selection on phenotypes. Paleobiologists revealed stasis and other phenomena that challenged evolutionary geneticists, and led to dialogue and deeper insights (Eldredge et al. 2005). Phylogenetic inference has provided ways of exploring problems ranging from character evolution to biogeography, from molecular evolution to the determinants of species diversity in communities.

An outstanding example of a question that will require a synthetic approach to answer is why adaptation in many instances is so rapid and fine-tuned, and why in other cases it lags or fails altogether. Despite the prevalence of local adaptation, despite the many cases of rapid adaptation in human-altered environments, despite the abundant heritable variation of most characteristics in most species, every species is limited to certain environments and has not broadened its niche to include others, and every species has a limited geographic range and has failed to adapt to the slightly different conditions beyond it. Phylogenetic studies have affirmed what taxonomists have long known, that many of the characteristics of organisms are phylogenetically conservative (i.e. they are synapomorphies), as are their ecological properties. Most temperate-zone species did not adapt *in situ* to the glacial-interglacial oscillations of climate during the Pleistocene; they shifted their geographic ranges by population extinction and colonization, and tracked their environment. And, of course, the vast majority of species that have existed have become extinct.

Why, if populations harbor so much variation, are they not more adaptable?

I suspect that deeper analysis of the genetic basis of adaptation will reveal that much genetic variation is not responsive to selection because of antagonistic pleiotropy or genetic correlations; that adaptations sometimes are based on rare combinations of new mutations; that evolutionary physiology will reveal complexities of form and function that slow or constrain change; that evolutionary ecology will show how adaptation to new environments requires evolutionary change not only in temperature tolerance, but in phenology and interactions with other species. Perhaps such insights will help us understand how to ameliorate the massive extinctions that will follow from the massive changes that humans are wreaking on the Earth.

Literature Cited

Blows, M. W., and A. A. Hoffmann. 2005. A reassessment of genetic limits to evolutionary change. Ecology 86:1371-1384.

de Jong, G. 2005. Evolution of phenotypic plasticity: patterns of plasticity and the emergence of ecotypes. New Phytol. 166:101-118.

Eldredge, N., and S. J. Gould. 1972. Punctuated equilibria: an alternative to phyletic gradualism. In T. J. M. Schopf (ed.), *Models in Paleobiology*, pp. 25-40. Freeman, Cooper, San Francisco.

Eldredge, N., J. N. Thompson, P. M. Brakefield, S. Gavrilets, D. Jablonski, J. B. C. Jackson, R. E. Lenski, B. S. Lieberman, M. A. McPeek, and W. Miller. 2005. The dynamics of evolutionary stasis. Paleobiology 31:133-145.

Forrest, B., and P. R. Gross. 2004. *Creationism's Trojan Horse: The Wedge of Intelligent Design*. Oxford University Press, New York.

Funk, D. J. 1998. Isolating a role for natural selection in speciation: host adaptation and sexual isolation in *Neochlamisus bebbianae* leaf beetles. Evolution 52:1744-1759.

Futuyma, D.J. 1970. Variation in genetic response to interspecific competition in laboratory populations of *Drosophila*. Am. Nat. 104:239-252.

Futuyma, D. J. 1979. *Evolutionary Biology*. Sinauer, Sunderland, Mass.

Futuyma, D.J. 1982. *Science on Trial: The Case for Evolution*. Pantheon, N.Y.

Futuyma, D. J. 1984. [Book review of] M.-W. Ho and P.T. Saunders (eds.), *Beyond Neo-Darwinism*. Science 226:532-533.

Futuyma, D. J. 1987. On the role of species in anagenesis. Amer. Nat. 130:217-226.

Futuyma, D. J. 1988. *Sturm und Drang* and the evolutionary synthesis. Evolution 42:217-226.

Futuyma, D. J. 2000. Potential evolution of host range in herbivorous insects. Pp. 42-53 in R. Van Driesche, T. Heard, A. McClay, and R. Reardon (ed.), *Host-specificity Testing of Exotic Arthropod Biological Control Agents*, Proc. X Internat. Congr. Biological Control of Weeds. USDA Forest Service, Morgantown, W.V.

Futuyma, D. J. 2008. Sympatric speciation: Norm or exception? Pp. 136-148 in *Specialization, Speciation, and Radiation: The Evolutionary Biology of Herbivorous Insects*, ed. K. J. Tilmon. University of California Press, Berkeley.

Futuyma, D.J., and G.C. Mayer. 1980. Non-allopatric speciation in animals. Syst. Zool. 29:254-271.

Futuyma, D. J., and T. E. Philippi. 1987. Genetic variation and covariation in responses to host plants by *Alsophila pometaria* (Lepidoptera: Geometridae). Evolution 41:269-279.

Futuyma, D. J., M. C. Keese, and D. J. Funk. 1995. Genetic constraints on macroevolution: The evolution of host affiliation in the leaf beetle genus *Ophraella*. Evolution 49:797-809.

Futuyma, D.J., and S.S. Wasserman. 1981. Food plant specialization and feeding efficiency in the tent caterpillars *Malacosoma disstria* Hübner and *M. americanum* (Fabricius). Ent. Exp. Appl. 30:106-110.

Futuyma, D. J. (ed.) and 16 others. 1998. *Evolution, Science, and Society*. (Document and executive summary on the web at www.amnat.org, executive summary in BioScience 49 (11) (1999), full document in Am. Nat. 158 (Suppl.):S1-S47 (2001).

Gould, S. J. 2002. *The Structure of Evolutionary Theory*. Harvard University Press, Cambridge, Mass.

Gould, S. J., and R. C. Lewontin. 1979. The spandrels of San Marco and the Panglossian paradigm: a critique of the adaptationist programme. Proc. R. Soc. Lond., B 205:581-598.

Kimura, M. 1983. *The Neutral Theory of Molecular Evolution*. Cambridge University Press, Cambridge.

Knowles, L. L. 2000. Tests of Pleistocene speciation in montane grasshoppers (genus *Melanoplus*) from the sky islands of western North America. Evolution 54:1337-1348.

Kuhn, T. S. 1962. *The Structure of Scientific Revolutions*. University of Chicago Press, Chicago.

Levins, R. 1968. *Evolution in Changing Environments: Some Theoretical Explorations*. Princeton Universitry Press, Princeton, N.J.

Lewontin, R. C. (ed.) 1968. *Population Biology and Evolution*. Syracuse University Press, Syracuse, N.Y.

MacArthur, R. H. 1961. Population effects of natural selection. Am. Nat. 95:195-199.

Mayr, E. 1962. *Animal Species and Evolution.* Harvard University Press, Cambridge, Mass.

Mayr, E., and D. Amadon. 1951. A classification of Recent birds. Amer. Museum. Novitates 1496:1-42.

Mayr, E., and W. B. Provine (eds.). 1980. *The Evolutionary Synthesis: Perspectives on the Unification of Biology.* Harvard University Press, Cambridge, Mass.

Mitter, C., and D.J. Futuyma. 1979. Population genetic consequences of feeding habits in some forest Lepidoptera. Genetics 92:1005-1021.

National Academy of Sciences Institute of Medicine. 2008. Science, Evolution, and Creationism. National Academies Press, Washington, D.C.

Wasserman, S.S., and D.J. Futuyma. 1981. Evolution of host plant utilization in laboratory populations of the southern cowpea weevil, *Callosobruchus maculatus* Fabricius (Coleoptera: Bruchidae). Evolution 35:605-617.

West-Eberhard, M. J. 2003. *Developmental Plasticity and Evolution.* Oxford University Press, Oxford.

Wilson, D. S. 1983. The group selection controversy: history and current status. Annu. Rev. Ecol. Syst. 14:159-181.

Wilson, D. S., and E. O. Wilson. 2007. Rethinking the theoretical foundations of sociobiology. Quart. Rev. Biol. 82:327-348.

6
Peter Godfrey-Smith

Professor of Philosophy
Harvard University, USA

---

1. Why were you initially drawn to discussions and research on evolution (or evolutionary aspects of your field)?

I reached evolutionary theory via the philosophy of mind. This might sound surprising, but quite a few philosophers of biology have this sort of history. In the 1980s when I was a student, philosophers were just starting to think seriously about using biological ideas to address the mind/body problem – or rather, they were starting to do this *again* after many decades of not doing so. There was discussion of using evolutionary ideas, in particular, in theories of meaning and representation. I was especially influenced by Fred Dretske's and Ruth Millikan's work. That started me thinking about evolution itself. Kim Sterelny, one of my teachers at this time at the University of Sydney, took a similar path. These interests expanded while I was in graduate school, at UC San Diego. I went there to work on both philosophy of mind and philosophy of biology, and was advised by Philip Kitcher. This was a period in which philosophy of biology was starting to become much more prominent within the philosophy of science, and there were dozens of interesting discussions arising. In 1990 I visited Dick Lewontin's lab at Harvard. Like many philosophers (including Bill Wimsatt, Elliott Sober, and Lisa Lloyd), visiting Lewontin's lab had a big influence on me, and from this point I began to look more closely at the structure of evolutionary theory. Then at Stanford I met Ben Kerr, who was at that time a graduate student, and we began collaborating on a series of papers about evolutionary models.

A feature of evolutionary theory over recent decades or so has been the development of several large-scale rival pictures of the biological world which are, in effect, mixtures of science and philosophy, even when the scientists involved do not think of their work as philosophical. One example of such a picture is Richard Dawkins' view, in which the notion of a "replicator" is seen as central to all our thinking about life and mind, and natural selection has primary importance in biology because it is the solution to the problem of explaining apparent design in nature. Evolution is a process in which replicators endlessly compete with each other, and the results include the appearance of a designed world. Another example is Richard Lewontin's view, in which evolutionary theory is first and foremost a theory of change, a theory which does great things within some domains but is easily over-extended and caricatured. Lewontin is skeptical about apparent design as a "special problem" for biology, and opposes a picture of evolution in which organisms are seen as mere products of their genes and as genetically adapted to an environment which "calls all the shots." In philosophy, Dennett has defended and extended a view like Dawkins'. Stephen Jay Gould's work also gives us another large-scale philosophy/biology mix of this kind, and developmental systems theory (eg., Susan Oyama) is another example. Views like this contain elements of science and philosophy mixed together in a complicated way. Sometimes the less empirical parts will be readily visible as metaphors, or as a separate commentary, but sometimes the mix is less obvious. These packages of views give us very different pictures of human agents, organisms, and our overall place in the world. I have spent a lot of time working on detailed topics within the philosophy of biology, but much of this work has been guided by a background interest in big rival pictures of this kind – pictures in which philosophical and biological ideas are tightly knitted together. I call this sort of work "philosophy of nature," as opposed to philosophy of science.

2. What does your work reveal about biological evolution (or evolutionary aspects of your field) that other academics, citizens, philosophers or biologists typically fail to appreciate?

In answering this question I will focus on some of my work on evolution itself, rather than work applying evolutionary ideas to the philosophy of mind.

First I will give some background. Darwin's central ideas are, of course, among the most important in the history of science. There is quite a long tradition of trying to give an abstract summary of what is essential to Darwinian evolution. (Here I am going to understand "Darwinian evolution" as synonymous with "evolution by natural selection," even though that is a simplification.) Darwin's own descriptions were mostly fairly concrete. He tried to work out how evolution works in actual organisms, here on earth. This was a strength of his work at the time; it was less speculative and more empirically grounded than earlier evolutionary writing. But quite quickly people began to try to give a more abstract description. Darwin was seen as identifying a sort of "schematic machine" that might operate in many systems. People tried to identify the core features of this machine, tried to distill out the essential features of the process he described. This would help us to better understand the central and familiar cases, and also help us to extend the theory to new domains.

There are two main ways of doing this, one older than the other. The older approach, which we can call "classical," says that, roughly speaking, any collection of entities will change by natural selection if three conditions are satisfied. There must be *variation* between the entities, some of the variants must have different *numbers of offspring*, and the variation must be *heritable* to some extent; offspring must tend to be more similar to their parents than they are to other individuals. A 1970 paper by Lewontin ("The Units of Selection," *Annual Review of Ecology and Systematics*) is the most common source for this summary. A second approach, developed by Richard Dawkins and David Hull, uses the concepts of a *replicator* and a *vehicle* (or interactor). According to this view, there are two roles played in any case of evolution by natural selection. One set of entities must faithfully pass on their structure in a copying process of some sort (replicators), and other entities (or the same ones again) must interact with their environment in such a way as to cause some replicators to copy themselves more than others.

I have looked closely at these summaries, with the aim of working out what their role is, which formulation is best, and so on. This makes it possible to re-organize some ideas in evolutionary theory in what I think is a productive way.

The first argument I make is that the "classical" view is superior. I see the replicator/vehicle analysis as a wrong turn. It is possible to have evolution by natural selection without replicators. The idea that there *have* to be some entities in an evolving system that are "copied faithfully" is a trap that has interesting connections to our psychology. It is easy for us to think about change in a system that is complex and "apparently designed" if we describe the change in terms of the activities of little agent-like things which persist and battle with each other over evolutionary time. So I think the right starting point is the classical view, but a lot can be done to improve the way this view is usually expressed and understood.

First, it is informative to look at these descriptions through a philosophical lens. A lot of standard summaries of how evolution works make idealizations – simplifying fictional assumptions – that have their roots in formal models in evolution but can be forgotten when giving verbal summaries. To pick a simple example, many formal models of evolution assume that everyone in a population reproduces at the same time and then dies (a "discrete generations" assumption). This simplifies the role of time in evolution. When thinking about the "fitness" of an individual, all one needs to worry about is how *many* offspring are produced, as opposed to *when* they are produced. But most organisms (for example, humans) do not have discrete generations. A lot of verbal descriptions of how evolution works are, in effect, verbal summaries of formal models, but sometimes the summary is given without recognizing the simplifications that helped make the formal model work.

The second point I'll make concerns a deeper issue. When people summarize how evolution by natural selection works, usually what is given is a set of necessary and sufficient conditions. Above I used the word "essence" a couple of times; people want to work out what is essential to Darwinian change. But a Darwinian context is one of gradients and intermediates. We should expect there to be lots of cases of systems which have a *partially* Darwinian character. And the really important or "paradigm" cases of Darwinian systems will have more features than those captured in the standard summaries. I have tried to build this picture into my treatment of what Darwinian change is. Standard summaries like the three-part Lewontin recipe discussed above give us a broad category of Darwinian change, but this category shades into "marginal" cases – cases where the Darwinian criteria are not clearly met but are approximated. Indeed, once we think about this we see that the clear, paradigm cases of systems that undergo Darwinian change must have evolved *from* marginal cases. Glenn Adelson calls this attitude "Darwinism about Darwinism" – as organisms evolve, the evolutionary processes they engage in also evolve.

When thinking about how this works, the concept of *reproduction* becomes important. Like Jim Griesemer, I think this is a concept that philosophers and foundationally-minded biologists have not given enough attention to so far. The history is something like this. The "classical" summaries of Darwinism, like Lewontin's, took the idea of reproduction for granted. We can see that in how the ideas of heredity and fitness were treated. All talk of heredity, and all talk of fitness differences in a population, presuppose that we know who is the parent of who. That makes it possible to count offspring, and also work out whether offspring resemble their parents. The replicator/interactor analysis (Dawkins and Hull) then came along, and rather than unpacking the idea of reproduction, they did things differently. Now the idea of "copying" or "passing on structure" was made central. It was assumed that all evolutionary processes would have replicators at work somewhere, and we don't have to worry about a general concept of reproduction so long as we can find the entities being faithfully copied. Once we see that the replicator view is not the right general analysis, we have to go back and deal with the questions that were left unanswered by the classical summaries. Then I think we find a lot of interesting new connections. In particular, reproduction itself comes in clearer and more marginal forms. The marginal cases are ones where reproduction is hard to distinguish from growth, or metamorphosis, or where the reproducing entities themselves do not seem to be real biological units. One way (not the only way) for a Darwinian process to be marginal is for the entities involved to exhibit only a marginal form of reproduction. Modes of reproduction themselves evolve, and take different forms in different parts of the tree of life. Evolutionary processes evolve as modes of reproduction evolve.

So a lot comes out of a re-examination and re-formulation of the classical summaries of Darwinian evolution. We get a somewhat different picture of the evolution process itself. The tree of life (which, we now know, is only sometimes tree-shaped) is a structure constructed from the births, lives, and deaths of organisms. But "birth," of course, is a word most appropriate for organisms roughly like us. Any collection of entities which can reproduce (and which varies, shows heredity, etc.) can undergo Darwinian evolution. These include integrated organisms with clear boundaries in space and time (like us, and octopuses) and also entities where it is hard to tell where one ends and another begins (corals, aspen). They include things like mitochondria and prions, which are small parts of organisms, and also various stretches of DNA. They include bee hives, other social groups, and symbiotic associations. As will be clear from some of these examples, this line of argument generates new ways of thinking about "units of selection" questions. It also helps us think about the idea of cultural evolution – this is another area where we find phenomena with a *partly* Darwinian character, and in this case cultural change can become more or less Darwinian on the rapid time-scale on which human habits of social learning and imitation change.

So starting from simple summaries, and rethinking and refining them, does take us somewhere interesting. I don't see these ideas as breaking dramatically with what other evolutionists have said. In many ways it is an attempt to start with widely-accepted and simple ideas, and think them through while avoiding wrong turns.

3. What, if any, practical and/or social-political and/or moral obligations follow from your work on evolution?

I am not sure what the answer to this is, and it is something I am thinking about at the moment. I doubt that "obligations" is the right word to use here, but I suspect that my view of evolution fits with some moral and political views than others. Both Ernst Mayr (in his famous discussions of "population thinking") and David Hull ("A Matter of Individuality") argued that there are moral consequences that follow from the kind of anti-essentialist thinking about biological kinds that Darwinism supports, and I've tried to follow up that style of anti-essentialist thinking (as has John Dupré). Mayr argued that a lot of harmful moral views make use of a "typological" way of thinking about human beings as a category, a view in which humans all share a form or essence of some kind. I think there is something right in Mayr's line of thought – I do think essentialist thinking tends to support moral outlooks I'd disapprove of, including various moral outlooks that value purity as a goal. On the other hand, it is hard to work out how the connections really work, once we get past obvious points that have to do with religious fundamentalism and the like. My own moral and political views are of a secular center-left kind. I don't think someone need be logically inconsistent if they support the same view of evolution as me, take a strongly anti-essentialist view about biological kinds, and then (for example) insist that only humans deserve any moral consideration at all, that abortion is always wrong, and the distribution of health care should be determined entirely by free markets.

It may be that the most we can expect in this area is something like this: views on the facts of evolution differ in how well they fit with various different "philosophies of nature." (Here I refer back to my answer to the first question.) A philosophy of nature contains a picture of how humans and their capacities are related to the rest of the living world and the universe at large. Each picture of that kind will tend to go naturally with some moral outlooks and not others. But "go naturally with" here does not mean anything like "imply." There is a lot of logical wriggle-room here, largely because of general issues about the relationships between facts and values. I'd like very much to say something more definite here, and I think this is a topic that we're not yet doing a very good job on.

4. What do you see as the most interesting criticism against your position in the biological or philosophical discussion of evolution?

Here is a line of criticism that puts together arguments made against me by various people, including Kim Sterelny and David Haig. In my discussion of evolution above, not much was said about genes. Genes are seen in that account as the main mechanism by which organisms and other entities inherit features from their parents, and also as entities that can reproduce, vary, (etc.) and hence engage in Darwinian processes in their own right. (Genes of course are also very important in metabolism, development, and just about everything else, but here I am focusing on their role in evolution.) So I start with abstract formulations of Darwinism that do not explicitly mention genes, and follow those formulations up as well as I can. But plenty of people would argue that it is *only* via focusing on genes that one can make any sense of evolutionary processes, in just about any real-world system. Whether or not one accepts a "gene's eye view" like Dawkins and Haig, one can regard change in gene frequencies as the very stuff of evolution, and describe change always in terms of genetic causes and effects. As I say, that is not how I did things above. And I can imagine someone arguing that this approach side-lines genes far too much. It might be true as a matter of abstract principle that you can have evolution by natural selection with any mechanism of inheritance at all – it might be true that genes are optional as a matter of logic. But (the objection continues), it is also true that once we move beyond Darwin's initial and very general statements, the only way we manage to get any kind of handle on evolutionary processes is by thinking about them in genetic terms. Fisher, Lewontin and others have argued that although Darwin did not *think* he needed anything like a Mendelian system as the basis of inheritance, *in fact* his theory did require this, for evolutionary theory to be able to explain anything significant. So I take seriously the argument that *as a matter of fact*, the best way to think about evolution here on earth is in overtly genetic terms, and the "abstract summaries" I start from and develop are only relevant as a simple first step.

I'll also mention a related and more general point. When a philosopher who does not work day-to-day in a laboratory setting comments on evolution, they naturally head towards general theoretical principles, and look hard at them and try to improve them. But it might be argued that in the actual business of sciences like biology, these abstract principles do not drive much that is important. What matters more is the development of detailed knowledge, including tacit "craft knowledge," of particular systems (and classes of such systems), through hands-on work. When a biologist with a very empirical orientation of that kind gives a general theoretical summary, either in verbal or mathematical terms, it might be argued that is not really meant to stand on its own. It needs to be interpreted in the light of tacit knowledge which someone who only ever works with general principles will not appreciate.

This argument might be directed against a biologist who only ever does abstract formal modeling, as well as at a philosopher. But I think it is particularly relevant to the sort of commentary on evolution that we see in philosophy of biology, and something that people who do that sort of work should worry about and keep an eye on.

7
Brian Goodwin

Scholar in Residence, Schumacher College, Dartington Hall, Devon, UK

---

1. Why were you initially drawn to discussions and research on evolution (or evolutionary aspects of your field)?

I always felt that the neo-darwinian theory of evolution in terms of genes and natural selection was seriously incomplete because it provided no explanation of the origins of the distinctive forms of different species of organism. By 'form' I mean morphology and behaviour, and these have to be generated during the developmental process before natural selection can act on them. Neo-darwinism effectively assumes that there is a continuity of forms within any species due to random variation in genotypes, and corresponding variation in phenotypes. Natural selection then 'chooses' the fitter forms to fit different habitats, and discontinuities between observed phenotypes in different species arise due to selection in significantly different habitats, with different selection pressures.

This does not fit with my understanding of development as a complex process that is intrinsically non-linear in its dynamics, which inevitably results in distinctly different dynamic attractors underlying distinct morphologies and behaviours as the general case. Complex dynamic systems like developing organisms generate emergent 'order for free', to use Stuart Kauffman's (1993) phrase, which means distinct forms that arise independently of natural selection. Of course once these forms have been generated, they have to be tested for their stability and resilience in their contexts, as is true for any real dynamical process. Hence natural selection has an important role to play in evolution as a dynamic stability testing process, but it is not responsible for producing the distinctive forms of living organisms in the first place. It accounts for the differential abundance of different species, not their origins. It is not an explanatory theory of the origin of species.

2. What does your work reveal about biological evolution (or evolutionary aspects of your field) that other academics, citizens, philosophers or biologists typically fail to appreciate?

It contributes to the tradition of Cuvier, William Bateson, Henri Bergson, D'Arcy Thompson and Waddington, among many others, that biological form reveals principles of coherence and dynamic order in the living state. Evolution is about a creative process that generates integrated wholes: organisms, communities, ecosystems, bioregions, Gaia, all reflecting their history and their intrinsic meaning, which for me signifies both their dynamic properties of resilience and adaptation (functional properties), and their intrinsic qualities of health and wholeness. My particular focus has been on the intrinsic dynamics of organismic development, understood as a process that is not determined by genes but by organisational principles that make use of genetic information in the generation of coherent wholes that fall into generic classes of biological form, all of which are transformations of one another. These classes are open, with unpredictable new patterns of order emerging during evolution.

3. What, if any, practical and/or social-political and/or moral obligations follow from your work on evolution?

The coherent states of organisms are revealed through their qualities such as health, wellbeing, and beauty, and the expression of feelings revealed in their behaviour, as well as through their functional, quantitative properties of exponential reproduction, diversity in ecosystems, and territorial behaviour. This gives rise to a science of qualities that can evaluate the conditions of experience of living processes, allowing us to observe the effects of our human interaction with other beings on the planet and to decide whether these are beneficial or detrimental. The result is an ethical science of skilful or right action by humans in relation to organisms, landscapes, ecosystems and Gaia as the coherent condition of the living earth, providing a basis for sustainable, compassionate, peaceful societies living in harmony with what we call nature.

4. What do you see as the most interesting criticism against your position in the biological or philosophical discussion of evolution?

I think the argument that I most respect is that there is no single perspective from which we can gain a comprehensive view of evolution. Rather, we need a multiplicity of interpretations, all of which are partially correct but also limited. This means that we must hold lightly to any interpretation, which is necessarily a story that presents a particular point of view, and avoid any literal reading of theory as Truth. Science is not about finding a metanarrative that is valid for all times and places, but about relevant stories that make meaning of bodies of evidence and provide appropriate courses of action, ethical as well as epistemological, within that cultural context and period of time. This is the ongoing story of continuous creation of the cosmos, which is the story that science keeps telling and retelling.

There is another widespread criticism of the perspective I adopt on evolution, which is the argument that human nature has been shaped by evolution to be competitive, aggressive, and dominated by self-interest, with societies largely functioning out of fear of scarcity. There are certainly elements of truth in this, but I consider that it is based on a simplistic neodarwinism that has become almost an ideology in its generality and power, especially as expressed in Western economic and capitalist activities.

5. With respect to present and future inquiry, how can the most important problems concerning evolutionary theory (or evolutionary aspects of your field) be identified and explored?

The most important problem at the moment for humans is whether we can survive in cooperation with the other beings of the planet, or if we allow our ideology to dominate our actions and eliminate ourselves in a wave of destruction of ecosystems and earth resources. Evolutionary theory needs to change in order to both understand more fully our past evolution, and to bring our future evolution into line with the principles that we can see operating in ecosystems and other species that provides for creative adaptation and resilience. Theory and practise need to converge in all our institutions so that they become consistent with new principles of coherent order that includes the whole of Gaia.

8
David L. Hull

Emeritus, Department of Philosophy
Northwestern University, USA

---

1. Why were you initially drawn to discussions and research on evolution (or evolutionary aspects of your field)?

I began my undergraduate work as a premed student at Illinois Wesleyan University in Bloomington, Illinois but soon became bored with all the memorization. Was I never going to learn anything? If at the time I had been taught even the rudiments of biological evolution, I would have applied to one of the graduate programs that taught evolutionary biology. As it is, I applied to a dozen or so graduate schools in a variety of fields and accepted the one that offered the most money – History and Philosophy of Science at Indiana University in Bloomington, Indiana. The goal of this newly emerging department was to teach philosophers of science a little history and historians of science a bit of philosophy. I was brought into the program because nearly all the work in history and philosophy of science dealt with physics, primarily relativity theory and quantum theory. It wouldn't hurt to have at least one biologist publishing in the history and philosophy of science.

While I was in graduate school, a controversy broke out in systematics – the classification of plants and animals – between traditional taxonomists and a newly emerging school of Numerical Taxonomy. Taxonomy seemed ideal for using computers. Nature is populated by millions of species, each with its own set of characteristics. These species in turn can be arranged hierarchically into more inclusive taxa – classes of organisms. Traditional taxonomists had to do all this work largely by hand. Computers could help bring increased order to this multiplicity. I must admit that I was not all that interested in systematics at the time, but I had two things going for me. First, right out of high school, I had worked for IBM in the army. I continued to work with computers at Wesley Memorial Hospital in Chicago while I was in graduate school. Second, in my undergraduate years, I had also learned a lot about a wide variety of plants and animals. Perhaps I could contribute something to this literature.

One of the most controversial positions in Numerical Taxonomy was an antipathy to theorizing. This "theorizing" took two forms: phylogeny reconstruction and evolutionary theory. Systematists use character distributions to reconstruct phylogenetic trees. However, numerical taxonomists were suspicious of the methods that traditional systematists used to reconstruct these trees. They seemed "circular." Numerical taxonomists were also suspicious of letting evolutionary theory "intrude" into classification. They found it equally "circular." Needless to say, I came down on the side of theories (Hull 1967). After all, it was the sort of issue that I was studying in my philosophy courses. As philosophers saw it, theory-free classifications are impossible and, if possible, undesirable. Finally, I was learning some theory – in this case evolutionary theory.

In all my naiveté I started submitting papers to journals, both philosophical and biological. In the manuscripts that I submitted to biological journals, I explained the relevant philosophy, and in the manuscripts that I submitted to philosophical journals, I explained the biology. I was not surprised that biologists took my work seriously – though I should have been. A graduate student in philosophy lecturing experts about their own specialties! But they did take me seriously, including the biggest names in the field – G. G. Simpson and Ernst Mayr. Not only was I accepted into the biological community but also I rapidly became part of the newly emerging field of the philosophy of biology. One of the first topics that arose in the philosophy of biology at the time was reduction in genetics. Just as Ernst Nagel (1961) had shown that thermo-dynamics was reducible to statistical mechanics, we were prepared to show that Mendelian genetics was reducible to molecular biology. The task turned out to be much more daunting than any of us expected, but I did learn an important lesson while working on this topic. Professional friendships really do matter. Those of us working on reduction in genetics were able to disagree with each other without coming to blows. Without realizing it, we were forming a research group. Little by little we were able to referee each other's work and get it published. Of these early philosophers of biology, Michael Ruse and I bonded.

In philosophy, one of the hot topics at the time was cluster analysis. Traditionally, the goal of classification was to find necessary and sufficient conditions that could define classes. Instead, in cluster analysis, all that is necessary is a class being defined in terms of enough characters weighted appropriately. Biological taxa (groups of organisms) seemed to be ideal examples of cluster concepts. For example, "Reptilia" can be characterized by a variety of traits, but other groups also possess these very same traits, and conversely, there are reptiles that lack them. If species evolve as gradually as biologists thought, then few if any can be defined in the traditional way. Instead they must be treated as cluster concepts (Hull 1965). Some biologists also extended this line of reasoning, not to species taxa, but to the species *category* itself.

2. What does your work reveal about biological evolution (or evolutionary aspects of your field) that other academics, citizens, philosophers or biologists typically fail to appreciate?

All of us are influenced by how and where we enter into our fields of study, an event that is largely ignored. Was I condemned to work in systematic philosophy for the rest of my life? Happily, not. A controversy over the levels of selection came to my rescue. The living world can be organized in two quite different ways. One concerns taxa. *Homo sapiens* as well as several other species all belong to a single genus – *Homo*. Homo and several other genera are included in the same Family, and so on. This is the traditional Linnaean hierarchy. Organisms are ordered in ever-more-inclusive taxa.

Living organisms can also be ordered in quite a different way as well – in an organizational hierarchy. At the bottom of this hierarchy are molecules that are organized into cell components, cells, tissues, and organs. Organs in turn make up organisms, some of which form super-organismal groups such as hives, demes and possibly even entire species. Do all organisms form entities at all these levels? No, but some do. The issue then is the level or levels of this organizational hierarchy at which selection can take place. In his classic paper, Dick Lewontin (1970) argued that Darwin's theory can be embodied in three principles:

1. Different individuals in a population have different morphologies, physiologies, and behaviors (phenotypic variation).

2. Different phenotyes have different rates of survival and reproduction in different environments (differential fitness).

3. There is a correlation between parents and offspring in the contribution to each in future generations (fitness is heritable).

According to Lewontin, selection can occur at a variety of levels from genes at one end of the spectrum to entire species at the other. The question is how often does selection actually occur at these various levels of organization? According to Lewontin, selection occurs primarily at the level of organisms, what he terms "individuals" (see also Keller 1999). At the height of this controversy, Richard Dawkins published his highly influential book, *The Selfish Gene* (1976). According to Dawkins, selection occurs only at the level of genes. It would seem that Lewontin and Dawkins disagree with each other profoundly. According to Lewontin, organisms are the primary, though not the exclusive, level at which selection can take place, while for Dawkins genes form the sole level at which selection occurs. But this disagreement turns out not to be as stark as it may seem.

Philosophers spend a lot of time disambiguating terms. Sometimes all this effort is worthwhile; sometimes not. The phrase "individual selection" is a good example of such ambiguity. Both "individual" and "selection" have multiple meanings. Throughout the literature on the levels of selection controversy, sometimes all that is meant by "individual" is "organism," but sometimes the term is also used in a generic sense to include such individuals as genes, cells (including gametes), kin groups, populations, species and communities. As harmless as alternating between the specific sense of "individual" and its more inclusive generic sense may seem, it has given rise to unnecessary confusion. All organisms are individuals, but not all individuals are organisms. When Lewontin says that "different individuals in a population have different morphologies, physiologies, and behaviors," he means organisms (see also Keller 1999). A second confusion stems from treating the term "selection" ambiguously.

In 1980 the editors of the *Annual Review of Ecology and Systematics* invited me to write a summary paper on individuality and selection. In this paper I argued that selection is not a single process but two – replication and environmental interaction:

1. Replicators are entities that pass on their structure largely intact.

2. Interactors are entities that directly interact as cohesive wholes with their environments in such a way that replication is differential.

3. Selection is a process in which the differential extinction and proliferation of interactors cause the differential perpetuation of the replicators that produced them.

Once this distinction between replication and environmental interaction is made, the disagreement between Lewontin and Dawkins is greatly reduced. Dawkins emphasizes the role of genes in replication. He is a gene replicationist. Lewontin, to the contrary, emphasizes the role of organisms in environmental interaction. He is an organism interactionist. And as Lloyd (1992) has argued at some length, the levels of selection controversy does not concern replication but environmental interaction, what Dawkins terms "vehicles." Although Dawkins and I mean much the same thing by "vehicles" and "interactors" respectively, we place very different emphasis on the two. Dawkins introduces the notion of "vehicles" but only to play down its importance, while I introduced the term "interactor" as a more perspicuous term for the process that makes replication differential (see Keller 1999). Lewontin and Dawkins really do disagree with each other, but these differences are exaggerated by terminological confusions. The term "selection" needs to be disambiguated. There are units of replication and units of environmental interaction but no units of selection. Selection does occur, but there are no units of selection.

No sooner was "individual" in the specific and generic senses distinguished than Michael Ghiselin (1974) extended the notion of individuality to include entities much higher in the organizational hierarchy than usual – including species. According to Ghiselin, species are best construed as individuals, not classes. When one examines the notions of species and organisms with sufficient care, the numbers of similarities between the two are striking. Both organisms and species have boundaries and locations in the literal sense. Biogeographers spend their careers ferreting out these boundaries. Both species and organisms have beginnings and endings in time. They can go extinct. Although lineages can both split and merge, merging seems to be more difficult than splitting. Some organisms exhibit considerable internal organization; species on the average do not, but the difference is one of degree, not kind.

The distinction between classes and individuals and where they belong in the organizational hierarchy may seem both esoteric and beside the point. What difference does it make? Reconceptualization is always a disconcerting exercise. One has to abandon or at least modify some very fundamental distinctions that have been around for a very long time. Species have been treated as the paradigm example of biological classes since the ancient Greeks. We ought to modify this notion only for very good reasons. The list is long. Can biological species be treated as essential natural kinds? Only with considerable care. Are there any biological laws that incorporate these putative natural kinds? Biologists have always been slightly embarrassed at their having so many natural kind terms (literally millions) and so few putative laws of nature that utilize them. If nothing else, treating species as individuals forces biologist to move up a level in the organizational hierarchy to look for anything that might count as a law of nature. Even if all swans were white, "all swans are white" would not be a very good candidate for a law of nature. Ornithologists are well aware that feather color in birds is highly variable

In looking back on the views that I introduced into the literature and shepherded down through the years, getting them accepted turned out to be quite difficult. We all recognize how hard discovery is, but we are not prepared for how difficult it is for these new views to become accepted or rejected for that matter. Because the issue of reduction in genetics fit so nicely into the views then current on reduction in physics, it rapidly became the received view, but later ideas were not so fortunate. Treating species as individuals was anything but standard. For centuries species were considered to be classes, and here Ghiselin and I were arguing that species have more of the characteristics of individuals than of classes. Biologists gave a fairer evaluation of this non-standard view than philosophers did, in part I suspect because they understood how radical this view was.

My analysis of "selection" was not so radical. I argued that selection was not one process but two. Asking for the units of replication and environmental interaction makes sense. As strange as it may seem, selection is inherently ambiguous. Evolutionary biologists continue to refer to the units of "selection" even after they see what the problem is. It also does not help that "species" is the same in the singular as in the plural. However, the most radical of my positions concerns sociobiology and the substitution of a general analysis of selection for reasoning by analogy, the topic of the next section.

3. What, if any, practical and/or social-political and or moral obligations follow from your work on evolution?

At roughly the same time as biologists were working on levels of selection and other issues in evolutionary biology, another even more emotional controversy arose – sociobiology. Because this controversy was so heated, I avoided it as best I could. It is one thing to argue about kin selection in flour beetles; it is quite another to argue for its role in human beings. Eventually, however, I did give way and published three papers on sociobiology that appeared in 1978 – the *Sociology of Sociobiology, Sociobiology: Scientific Bandwagon or Traveling Medicine Show,* and *Altruism in Science: A Sociobiological Model of Cooperative Behavior among Scientists.* In these papers I evaluated sociobiology to see if this new research program came up to the standards of genuine science. My conclusion was that, yes, it did, given reasonable criteria for what count as genuine science. But more importantly, I used scientists themselves as my research organisms. If people in general can behave altruistically under certain conditions, then so can scientists – including me. If, to the contrary, people in general cannot behave altruistically, then how about scientists? And here is yet another instance in which the role of professional friendships can influence the course of science. Ruse was not in the least interested in anything like a general analysis of selection. We still cooperated.

One other characteristic of these three papers is worth noting. They appeared in journals that my fellow philosophers of science were not likely to read – *Society, Animal Behavior,* and the *New Scientist.* I published these papers where I did because I wanted to stake out priority without my fellow students of science being any the wiser. Priority over what? The idea of dealing with the role of "kin selection" in the social structure of science. The relevant "kin" are not *biological* kin but my fellow scientists. Scientists are altruistic in their relations with other scientists because it is in their own best self-interest to do so. I wanted to keep these ideas confidential because I suspected that they were too radical for my fellow philosophers. "Altruism" explained in terms of self-interest might rub them the wrong way. Eventually in 1988 I did come out when I published a heavy-duty book entitled *Science as a Process: An Evolutionary count of the Social and Conceptual Development of Science.* This book was followed by a contribution to the *Behavioral and Brain Sciences* in 2001 in collaboration with Rod Langman, an immunologist, and Sigrid Glenn, a behavioral psychologist.

One of the most controversial topics in Darwin's formulation of natural selection was the nature of what we now term "genes" and the roles that they play in evolution. To use present-day terminology, in gene-based biological evolution, genes are the primary replicators. Although advocates of sociobiology tend to favor gene replication, there is no necessary connection between sociobiology and genes being the units of replication. Organisms would do almost as well. A second topic that played a major role in Darwin's theory is the distinction between natural and artificial selection. In natural selection, "nature" does the selecting. In artificial selection plant and animal breeders do the selecting. Among the usual sheep, a short-legged mutant cropped up. If left to "nature," this sheep and its kin would soon die off. However, human beings selected this sheep to breed in order to establish populations of what came to be known ancon sheep. At this stage plant and animal breeders could not generate desired mutants. All they could do was favor those that cropped up all on their own. Today we can actually produce the mutants that we want. Hence, artificial selection is doubly "artificial."

Evolutionary epistemology is one of the earliest efforts to apply evolutionary biology to conceptual change. How come flying squirrels can judge distances so well? The answer is that if they could not, they would not have survived. How come people are so good at recognizing human faces? Such an ability played and continues to play an important role in human interactions. In the early days of evolutionary epistemology, the line of argumentation was to reason from strictly biological units to sociocultural evolution – what we now term "memes" (Dawkins 1976, Distin 2005). On this interpretation, both genes and memes function as replicators, and neither is taken to be paradigm. Instead of reasoning from genes to memes, the goal is to provide a general analysis of selection that handle all sorts of selection from gene-based biological selection, the reaction of the immune system to antigens, as well as individual learning and the growth of languages (Hull, Langman and Glenn 2001).

This issue is fundamental. From the beginning, evolutionary epistemologists such as Donald Campbell (1974) worked simultaneously on two sorts of evolutionary epistemology. One sort is reasoning by analogy from gene-based replication to other sorts of replication. In this inference, genes are primary. On the second sort of evolutionary epistemology, all sorts of replication, both genes and memes, are equally "replicators." The goal is to provide a general analysis of selection (Hodgson et al 2008).

The major source of discontent with sociobiology is the implications that it might have for the welfare of human beings. Michael

Ghiselin (1974), E. O. Wilson (1975) and Richard Dawkins (1976) brought out their major works right at the time that opponents of the Vietnam War were using the techniques of social constructivism to criticize this stupid and evil conflagration. Although these authors were not condemned for actually contributing to this war in a direct manner, they were criticized for doing so unwittingly. The main assumption of much of this literature was that reason, argument and evidence matter. Academics are morally obligated to bring their training to bear on socio-political controversies. We should not sit idly by as a great evil engulfs us.

One of the problems with academics getting involved in socio-political causes is deciding which causes. Looking back at the American Civil War, I hope that if I lived back then, I would have worked to free the slaves, but perhaps not. Who knows? I opposed the Vietnam War, but looking back, I did not do as much as I should have done to oppose it. Of course, what can academics actually do? One controversy to witch evolutionary biologists were able to make a major contribution – Creationism. We understood the biology and, even more importantly, we had a professional understanding of what counts as science. I wish that I could say that I did my share in keeping Bible stories out of biology classes, but I can't. Once again Michael Ruse put himself on the line. He also took on the clash between science and religion. Perhaps, if I had been raised a Quaker, I might not have exhibited such antipathy to religion, but Sister Mary Dorothy took care of that for me all on her own. But I have rigged the preceding examples of socio-political and moral obligations. They exhibit the usual good guy/bad guy syndrome. How about an example that is currently undecided? Right now the treatment of undocumented aliens in the United States is a real issue. Who is right?

I tend to be cynical about how much of a difference reason, argument and evidence make in the resolution of social ills. Propaganda, social pressure and brute force seem to bear most of the weight. How about the publication of a finely tuned argument? At times I am led to conclude that a good argument has never convinced anyone. Hitler proclaimed Germans to be the master race. One might think that he would modify his claim in the face of a black runner winning four gold medals in the German Olympics. Would anything have changed if Owens had come in second? Data supposedly matter. Sometimes, when all else fails, maybe they do.

Most examples of the interrelations between science and society paint a very ugly picture of science and society. Supposedly Darwinism promotes euthanasia, abortion, eugenics, homosexuality, Nazism and what have you. It doesn't help that I have nothing against any of these supposed "ills" save Nazism. However, how about variability? In general, people do not like variability all that much, but one of the main messages of evolutionary theory is that variability is good. Without it we would soon go extinct. The more varied the human race is, the more likely it is that we will not go extinct. Variability has its limits, but thus far we are not close to approaching them.

Scientists have also pointed out other equally "practical" obligations. For a very long time, evolutionary biologists have warned our politicians of the dangers of our over-use of antibiotics. Without realizing it, we were breeding drug resistant viruses and bacteria. We now have to live with the results. The impact of climate change is yet another example of unheeded obligations.

Thus far I have not touched on morally sensitive issues such as homosexuality. One would think that a behavior that decreases the likelihood of passing on one's genes would result in fewer of these genes being passed on, but non-reproductives are commonplace, especially among social insects. Perhaps the explanations suggested for these organisms might work for human beings. How can adaptations become established if the organisms who possess them never reproduce? One explanation is kin selection. Jesuit priests may never reproduce themselves, but they might help out their siblings and pass on their genes indirectly through them. Gay people tend to like this explanation because under it gay people serve a function. Sticking around the nest to help one's siblings may not be all that romantic, but it is better than the usual explanations suggested for gay people. Kin selection may be the correct explanation for people preferring members of their own sex for sexual and romantic love. How many homophobic religious leaders have thereby been persuaded to give up their views on gay people burning forever in hell? Not many I suspect.

Sexual dimorphism is another common phenomena in the living world. In some species, males and females differ from each other. In some species females are larger than males; in others males are bigger. The human species happens to be one of those sexually dimorphic species. Males and females differ from each other to some extent. Here the implications are fairly direct and not to the liking of quite a few people. Developing a unisex society is going to be quite difficult to provide and require constant effort if we do. Then there is the issue of intersexes. One or more of the evolutionary explanations provided for intersexes in other organisms might apply to people as well.

4. What do you see as the most interesting criticism against your position in the biological or philosophical discussion of evolution?

In my early work on systematics, I tended to side with those systematists who saw a central role for "theories" of various sorts in their work, operational problems be damned. I now appreciate the work that scientists put in trying to show how to apply these terms. In my early work on reduction I was not prepared for how malleable scientific theories actually are. If there were one and only one formulation of Mendelian genetics and one and only one formulation of molecular biology, then reducing one to the other might be feasible, but such is not the case. No sooner do we acknowledge a problem and its possible solutions than another one crops up. Pretty soon none of these revamped theories look much like the original theories. Historians of science frequently decry how cavalier philosophers are with respect to history of science. They take offense at calls to rewrite history so that it fits some philosophical position. As the years have gone by, I have come to appreciate "history for its own sake," as impossible as it might be.

One thing that analytic philosophers do is to set forth a version of a particular word or phrase and clean it up. Others study this analysis to find various shortcomings and produce a refined version. Then another round of revisions is introduced. Eventually, we reach a point of no return and switch to another term. In my analysis of "selection" I define "selection" in terms of replication and environmental interaction. However, I do not include any reference to variability – an element that is absolutely necessary for selection to occur. I discussed it but did not include it in my definition. In addition, I refer to replicators "producing" interactors. "Producing" may be appropriate for the relation between genes and phenotypic traits but not for conceptual change. In general, lots more work needs to be done on treating conceptual change as at least partially a selection process.

5. With respect to present and future enquiry, how can the most important problems concerning evolutionary theory (or evolutionary aspects of your field) be identified and explored?

I have only the vaguest idea about the future of evolutionary theory. We have gone through an exciting period in evolutionary biology. Will it continue? We now understand fairly well the implications of molecular biology for evolution. Embryologists are making headway in their field as well. Will the result be yet "another new synthesis"? It all depends. One thing is certain. We must make room for the next generation.

References

Campbell, D. T., 1974, Evolutionary Epistemology, In P. A. Schilpp (ed.), *The Philosophy of Karl R. Popper*, LaSalle, IL.: Open Court Publishers.

Dawkins, R., 1976, *The Selfish Gene*, Oxford: Oxford University Press.

Dustin, Kate, 2005, *The Selfish Meme*, Cambridge: Cambridge University Press.

Ghiselin, M. T., 1974, *The Economy of Nature and the Evolution of Sex*, University of California Press.

Hodgson, Geoffrey M., David L. Hull, Thornbjorn Knudsen, Joel Mokyr and Viktor J. Vanberg, 2008, In Defense of Generalized Darwinism, *Journal of Evolutionary Economics*, forthcoming.

Hull, David L., 1965, The Effect of Essentialism on Taxonomy: Two Thousand Years of Stasis, *The British Journal for the Philosophy of Science*, 15:314-326.

Hull, David L., 1967, Certainty and Circularity in Evolutionary Taxonomy, *Evolution* 21:174-189.

Hull, David L., 1988, *Science as Process: : An Evolutionary Account of the Social and Conceptual Development of Science*, Chicago: University of Chicago Press.

Hull, David L., Rodney E. Langman, and Sigrid S. Glenn, 2001, A General Account of Selection: Biology, Immunology and Behavior, *Behavioral and Brain Sciences*, 24: 511-573.

Keller. Laurent (ed.). 1999, *Levels of Selection in Evolution*, Princeton University Press, NJ.

Lewontin, R. C., 1970, The Units of Selection, *Annual Review of Ecology and Systematics*, 1:1-18.

Lloyd, E. A., 1992, Units of Selection, *Key Words in Evolutionary Biology*, E. F. Keller and E. A. Lloyd (eds.), Harvard: Harvard University Press,

Nagel, Ernst, 1961, *The Structure of Science; Mechanistic Explanation and Organismic Biology*, New York: Harcourt, Brace & World.

Wilson, E. O., 1975, *Sociobiology: the New Synthesis*, Cambridge: Harvard University Press.

9
Eva Jablonka

Professor

The Cohn institute for the History and Philosophy of Science and Ideas, Tel-Aviv University, Israel

---

1. Why were you initially drawn to discussions and research on evolution (or evolutionary aspects of your field)?

My work over more than 20 years has been on evolution and the role of epigenetic inheritance – the inheritance of heritable variations that do not stem from variations in DNA base sequence. And because I am an evolutionary biologist, I am answering this question by presenting a historical account of how my involvement with evolutionary biology unfolded.

I was drawn to biology *through* my interest in evolutionary biology. As a child I loved animals and plants (as things of beauty and as companions, not as objects of scientific inquiry). My scientific interest in biology and evolution, however, started when I was 17. Then I read Darwin's *Origin* and Koestler's *The Ghost in the Machine*, and the philosophical and social questions that these books raised fascinated me and linked the aesthetic and emotional pleasure in the living world with an intellectual pleasure. The narrative structure of evolutionary explanations, the storytelling aspect of it, so evident in Darwin's *Origin*, also attracted me. Another attraction of evolutionary biology was the controversy about neo-Darwinism that Koestler presented in his book – the arguments against the possibility of the inheritance of acquired characters, which he defended. I found this debate exciting and challenging, and was drawn to Koestler's rebellious position. Darwin's *Origin* convinced me of the explanatory power of natural selection – the idea made wonderful sense to me. I did not, however, understand why biologists were so opposed to the idea of the inheritance of acquired characters. Darwin did accept it, and there was no reason to think that natural selection could not lead to the evolution of "intelligent" biological systems capable of generating better-than-blind genetic variations in the right conditions. Foresight, after all, is based on the memory of past experience, so it was not clear to me why it was impossible to evolve such memory. My point of departure was not to look for an alternative to natural selection as an explanation of functional complexity. Rather, it was *because* I was so convinced of the power of natural selection that I thought that there must be evolved mechanisms that lead to the generation of some "acquired" characters.

I started studying biology when I was 20. As a freshman, my first essay on a self-chosen topic was on the inheritance of acquired characters, and on that occasion I discovered Waddington (he was mentioned by Koestler, and that is how I came to know about him). It was a great discovery for me, and it made me realize that it is possible to shift from acquired to "innate" characters without any need for a direct feedback between soma and germline. This was in the very early 1970s, and at the time I knew of no respectable examples for the inheritance of acquired characters. I thought that Waddington had solved the problem. He gave such a beautiful answer in terms of what was known of heredity in that period (of which I had only a vague idea, as yet having taken no genetics course). I still had some doubts, but I had no clue about how else the question could be approached. I had no clear research direction, just a feeling of excitement about evolution, curiosity, and a rebellious sprit.

I went to study biology in Birkbeck College in the University of London in 1973, when I was 21 years old. I spent only one year there, but it decided my future. There I met Marion Lamb, who was my teacher of genetics. Marion was a superb teacher; she challenged the students and made us very excited about genetics. When I first talked with her, after a few lectures of genetics, I enthusiastically mentioned Waddington, and Marion who was, as I later learnt, a great Waddingtonian and taught his ideas in advanced courses, told me that I should "learn to walk before I start to run". So I did learn. I studied genetics, and later did my PhD degree on this subject. Marion and I became close friends, and although it took many years until we started working together and developed our views on epigenetic inheritance and its role in evolution, we always talked about evolutionary problems. Marion sent me books on evolution as they were published in England and before they reached Israel – books like Dawkins' *The Selfish Gene* and Wilson's *Sociobiology* – and we discussed the ideas and the problems they presented.

There was no biological problem which Marion would not discuss from an evolutionary perspective. She came from the British school of biology, and did her PhD under the supervision of John Maynard Smith on the genetics of aging, so she framed biological problems naturally within an evolutionary framework. She also studied with Medawar and was open to the possibility that there may be good biological mechanisms for the inheritance of acquired characters. However, at that time, there were other pressing biological problems that had priority for us: sociobiology and the social and philosophical difficulties that it presented, the debate between the neutralists and the selectionists, punctuated equilibria, and the relationships between ontogeny and phylogeny that were explored in Gould's 1977 book. Marion's influence made me think about biological problems, including those in which I became involved as I continued my studies, within an evolutionary framework. It was natural to ask about the evolutionary implications of the processes or systems I studied, and about their own evolution.

I did my Master's degree in microbiology, and when I thought of continuing for a PhD, I wanted to work on mechanisms generating genetic mutations in microorganisms, and to investigate the possibility that the mechanisms that generate mutations are linked to regulatory mechanisms of transcriptional activation. But it was not possible to find a lab that was interested in this, so I started my PhD on the tantalizing problem of X chromosome inactivation in female mammals, in Menashe Marcus' lab in the Genetics department in the Hebrew University in Jerusalem, under his and Howard Cedar's joint supervision. I was fascinated by the fact that the state of a whole chromosome can be regulated as a unit, and looked at the relation of DNA methylation and X chromosome inactivation, which was one of the problems that Marcus, as well as Cedar and Aaron Razin, were working on. The focus on DNA methylation, which is a chromatin marking system that is involved in the regulation of transcription but is also a cell memory mechanism, led to Robin Holliday's papers and to thinking about DNA methylation as a cellular memory and a heredity system.

Marion and I discussed the evolutionary aspects of X inactivation, but it took a few years for us to realize the transgenerational, evolutionary potential of variations in DNA methylation, and of other cell heredity systems that generate epigenetic variations. In 1986 we started to work together in a formal way, that is, writing papers together and not just discussing evolution for hours over a cup of coffee (for me) and glass of whiskey (for Marion). Our first papers were on the evolution of X-chromosomes, and our starting point was to find the common denominators of sex-chromosome regulation in different taxa and under different developmental conditions. In 1987 I went for a short, four-month, post-Doc in Anne McClaren's lab in London to work on X-reactivation/inactivation during female meiosis and in early ontogeny. There, I briefly met Robin Holliday, whose work I admired. Marion and I read his 1987 paper on the inheritance of epigenetic defects, and it reinforced the ideas that we were developing on the transmission of epigenetic variations between generations and about the evolutionary importance of such inheritance. We then finished writing our first paper on epigenetic inheritance, "The inheritance of acquired epigenetic variations". The paper focused on DNA methylation, and since DNA methylation is involved in both regulation and cell heredity, we argued for the possibility of soft "Lamarckian" inheritance, the between-generation inheritance of developmentally induced and regulated epigenetic variations. The paper was rejected by Nature and Science (editorially) and was published a year and half later, in 1989, in the Journal of Theoretical Biology.

By the 1990s we had written several papers about the role of epigenetic inheritance in evolution, expanded our notion of epigenetic inheritance to include self-sustaining loops and structural three-dimensional templating mechanisms (we called the different factors and mechanisms epigenetic inheritance systems or EISs, an acronym suggested by John Maynard Smith), discussed the role of epigenetic inheritance in speciation, adaptive evolution, the evolution of genomic imprinting, aging and the evolution of individuality, and considered the adaptive evolution of epigenetic inheritance strategies. We had difficulties in publishing most of these papers. In one case it took 4 years to get a paper published, and another paper never saw the light of day. As a result of these troubles, we decided that it would be easier to write a book – it seemed to us that we are spending a lot of energy fighting editors and saying bits and pieces, while we really have a very clear and comprehensive idea of what we want to say. We had a notebook with ideas we jotted down, and planned a book. Our 1990 book-plan is exactly what our 1995 book *Epigenetic Inheritance and Evolution: The Lamarckian Dimension* was about, in terms of both content and structure.

We were fortunate, because we had the support of John Maynard Smith. We are sure that we would not have been able to publish our first paper and our first book without his active intervention. His opinion was sought and he was very positive. He was interested in our ideas although he generally disagreed with us. We argued with him privately and publically (in the pages of *Journal of Theoretical Biology*). It was always productive and thoroughly enjoyable, often over a bottle of wine that he and Marion shared.

Our interest in epigenetic inheritance grew into a more general interest in non-genetic systems of transmission of information and their roles in, and implications for, evolutionary biology. My 2000 book with Eytan Avital, *Animal Traditions*, explored the evolutionary implications of transmission via social learning in animals, and my papers with Daniel Dor examine the co-evolution of genes and linguistic culture for the evolution of the human capacity for language. Marion and I extended and synthesized this general epigenetic approach in our most recent (2005) book, *Evolution in Four Dimensions*.

I am now working with Simona Ginsburg on the biological basis of the ability to experience (feeling thirsty, which a dog or a human may experience, rather than having sensory states indicating a lack of water, as in plants). We approach this problem from the evolutionary end, trying to figure out when the transition to experiencing occurred in animal evolution, how it can be identified, and, if identified, how it can be studied. We are still at the beginning of this intellectual adventure, so in the rest of this interview I shall address only epigenetics and epigenetic inheritance.

2. What does your work reveal about biological evolution (or evolutionary aspects of your field) that other academics, citizens, philosophers or biologists typically fail to appreciate?

What people appreciate or fail to appreciate is (fortunately, in our case) changing over time. Our evolutionary-developmental approach is a case in point, although how opinions changed and what is still thought of as a problem, depend on the type of epigenetic inheritance one addresses. Epigenetic inheritance encompasses both transmission that bypasses the germline altogether (soma-to-soma transmission, transmission through various developmental feedback loops, through social learning in animals, and through symbolic communication in humans), as well as epigenetic inheritance, where the single cell is the unit of transmission (cellular epigenetic inheritance). Most people accept that cultural evolution in humans is very important, but they tend to think about it as an independent axis of transmission and evolution, and not as a central factor which drove hominid genetic evolution (which is the opinion that Marion Lamb, Eytan Avital, Daniel Dor and I hold). There is, however, a growing appreciation of the interactions between genetic and cultural human evolution, so the guiding role of cultural inheritance in human evolution is beginning to be widely recognized. Cultural evolution through social learning in non-human animals is also getting a lot of attention at present, but there is still great controversy about the ubiquity of such transmission in animals, and the evolutionary importance of animal traditions. Eytan Avital and I think that animal traditions are very common and drove many important aspects of animal evolution, mostly (although not exclusively) the evolution of their behaviour. However the main difficulty for the epigenetic approach is cellular epigenetic inheritance, which is found in all taxa from bacteria to man, and allows the transmission of epigenetic variations through the germline.

When Marion Lamb and I first started working on cellular epigenetic inheritance in the 1980s, there was a lot of general scepticism about the very fact of epigenetic inheritance. Biologists just did not believe that it happened. A bit later there was reluctant acceptance; people did admit that there are indeed some cases, but these cases were assumed to be exceptional, unstable, and of no general importance. However, in the mid-1990s, epigenetics, the study of the processes that underlie developmental flexibility and stability, became a very hot topic. Molecular biologists described more and more factors involved in the chromatin marking EIS, which includes DNA methylation, modifications of the histones around which DNA is wrapped, and DNA-binding non-histone proteins. The intricate regulatory interactions between these factors were being explored, and there was evidence for their transmission between cells through mitosis and sometimes through meiosis. A new and amazing regulatory and cell heredity system, the RNAi system, was also discovered, and it began to be intensely explored at the end of the 1990s. The process is ongoing; new surprises are constantly emerging, and old puzzles, such as some very odd patterns of inheritance in ciliates, are at long last getting a good mechanistic explanation. As these studies were being conducted, it became clear that epigenetic control mechanisms and epigenetic variations had important implications for medicine and agriculture, and that they are of central importance in development. Cases of transgenerational epigenetic inheritance, especially in mammals, got some general attention and were described in popular science journals like the *New Scientist*, or the *American Scientist*. The only area of biology where the importance of epigenetic inheritance is downplayed is evolutionary biology.

What is the problem? Epigenetic inheritance, the inheritance of heritable variations that do not stem from variations in DNA base sequence, is problematic because it legitimizes soft inheritance. Soft inheritance occurs when new variations that are the result of environmental effects, use and disuse, or other factors are transmitted to the next generation. Soft inheritance was considered a heresy in biology for quite a long time. It was "Lamarckian", and any kind of Lamarckian inheritance – the inheritance of acquired/induced characters – was assumed to be wrong. The anti-Lamarckian conviction, which was one of the defining assumptions of the Modern Synthesis that was forged in the Western world from the late 1930s onwards, was reinforced by the cold war. Lamarckism became associated with Soviet Lysenkoism. Lysenko, a tyrant scientific leader in the totalitarian Soviet regime, denied Mendelian genetics and persecuted Soviet geneticists who adhered to Mendelism. So, in the West, Lamarckism became associated with Lysenko and with bad biology, and anything that smelled even remotely of Lamarckism was very suspect. Since all the epigenetic control mechanisms which underlie the transmission of epigenetic variation from one cell generation to the next (when the cell is the unit of transmission) obviously permit, and in many cases necessitate, soft (Lamarckian) inheritance, it was very difficult to accommodate within the Modern Synthesis. Moreover, accepting soft inheritance requires a kind of gestalt switch in the way one thinks about heredity and evolution, with development, and developmental units, rather than the genes, having explanatory primacy. According to the developmental view, the genes are seen as heritable resources for development rather than as prime movers. The question about what other resources contribute to development and to heredity is therefore legitimate. However, the hegemony of the gene-centred view over the last 30 years means that the development-oriented approach offered by the epigenetic perspective, is clashing with the hegemonic view, and therefore has to counter the resistance of conceptual inertia.

Accepting soft inheritance also has implications for the modelling of evolution: models of evolution need to accommodate induction, acknowledge that "mutation" pressure (more generally, "heritable variation" pressure) and selection coefficient are not independent, take into account the way in which epimutations appear in populations (affecting many individuals simultaneously and being able to spread horizontally among individuals), and recognize the possibility of coordinated, multiple, epigenetic changes in the same individuals, especially under conditions of genomic or ecological stresses. If these were included, the result would be very different models of evolutionary change. It is not surprising that there is some reluctance among people socialized in and working within the 'traditional' theoretical framework to make the necessary changes.

3. What, if any, practical and/or social-political and/or moral obligations follow from your work on evolution?

As I see it, the practical and moral obligations are connected: I believe that human beings have the obligation to change conditions of poverty and injustice if they have the practical means to do so. However, the obligations that follow from the work on epigenetic inheritance and the development-oriented view of evolution are not derived from the work per se, or only from its practical implications, but first and foremost from ethical and social/political convictions, and especially from a standpoint about social responsibility. One important facet of scientific work is social: scientists work with colleagues, in a society, in a political milieu. They are not working and living in an ivory tower, nor in a debating club. They do their highly privileged work in a painful reality, full of social injustice, where a lot of people are starving, suffering from aggression and inflicting aggression on others, and polluting the world around them. Scientists, like other people, have, of course, different ways of coping with the problems of the world, depending on their views. My own view is fundamentally socialist. I believe that economic inequalities have to be remedied through the intervention of both individuals and social groups (e.g. state, community), with the privileged members of a society, as well as privileged societies (the "first" world) investing in the remedy of economic, social and political inequality of less privileged individuals and communities. I think that what the work on epigenetic inheritance is showing is that many of the diseases and conditions that we thought were genetic "fate", that were "in the genes", are in fact environmentally induced, and much of the suffering involved can be remedied if we counter and reverse the effects of maladaptive, sometimes trans-generational, gene expression patterns. This means that we have more social responsibility than we thought, and we have to exercise it.

The practical implications of epigenetic inheritance are numerous: the research is already impacting almost every area of medicine, such as cancer research, the study of chronic diseases, environmentally induced diseases, aging, and epidemiological studies. For example, it seems that in some cancers (for example, some colon cancers) cell-heritable alterations in chromatin, such as alterations in DNA methylation, initiate the process of cancerous transformation, with alterations in DNA sequences following suite. Similarly, epigenetic modifications are implicated in some chronic diseases, in environmentally induced diseases, and in ageing. What happens during early development, sometimes even in the uterus, may have long term and possibly heritable effects, so a new branch of medicine, developmental medicine, is becoming established. We must also take on board the fact that pathogens may have epigenetic ways of countering our defences against them, so we have to develop medicines that will neutralize such variation-generating mechanisms.

Since we now realize that the conditions of development are critical for human well-being, the quality of the environment is crucial. For example, since it seems that starvation, occurring during sensitive periods of development, has trans-generational effects, not only do we have to prevent starvation, but we also have to understand the long-term effects that it has, and to develop measures to counter them. The same is true for the obesity epidemics, and possibly also for the effects of drugs and for some psychological stresses, since studies in mice and rats suggest that the effects of such stresses, when chronic or when occurring during "sensitive" developmental periods, may linger for several generations. Pollutants, as we know, have many developmental effects, and some of them seem to persist for generations. For example, it seems that the androgen suppressors that are present in a widely used fungicide, when administered to a pregnant rat, can have detrimental effects in the following three generations of her offspring.

Recognizing the epigenetic dimension of diseases give us a double responsibility. First, we need to do what we can to prevent the conditions that lead to these diseases. This means involvement in public health programs, and taking a long-term, developmental view of public health intervention polices, in addition to attempting to ameliorate immediate, short-term problems. Second, it means that new types of medications, which make use of the knowledge of epigenetic factors and mechanisms, need to be developed. This is already happening. Medications based on small regulatory RNAs are being produced; chemicals that alter DNA methylation and other components of chromatin are being developed; and the epigenetic state of tissue chromatin (e.g. DNA methylation patterns) is beginning to be used for diagnosing the stage of disease development. Another important consideration, which is relevant for both medicine and agriculture, is that the cloning of animals demands that we take into account epigenetic re-programming. Full epigenetic re-programming is required if cloning is to result in healthy tissues or individuals. Great caution must therefore be exercised, and a lot more knowledge needs to be gathered, if cloning techniques are to be implemented. Needless to say, this is especially pertinent in the human case.

There are other problems that a focus on non-genetic information transmission alerts us to. Conservation policies have to take into account the fact that what needs to be conserved and reconstructed depends not only on genetic information but also on information that is transmitted at the supra-genetic level. For example, animals reared for generations in zoos cannot just be liberated into the wild with the assumption that their genes will "tell them what to do", because they will not be able to cope – they have no socially-transmitted information about how to deal with natural foods, predators, and competitors. Another related aspect has to do with our destruction of bio-diversity on our planet. Since epigenetic variations are sometimes dependent on persistent interactions among different biological species, freezing the genetic material of endangered species may not be sufficient to conserve them. We have to realize that the destruction of ecological niches, with all the historical epigenetic "memories" that they carry, is really irreversible. This can give us another strong incentive to counter the massive destruction that we are bringing about.

I have not exhausted all the implications, either practical or moral, of epigenetic inheritance – this is quite impossible, because there is no domain of the biological sciences which is not involved. The general message is that since the epigenetic perspective stresses environmental effects as crucial inputs for both individual development and for heredity, the quality of the environment, for individuals and communities, has to be a major part of the research agenda of socially-informed biological practices. In a very literal sense, the sins of past social injustice are visited upon us, while our own social and ecological sins will be visited, with a vengeance, on our descendants.

4. What do you see as the most interesting criticism against your position in the biological or philosophical discussion of evolution?

The position I presented hinges on the conjecture that epigenetic inheritance is important and common, and that epigenetic variations cannot be understood only in terms of genetic (DNA) variations, so there is a partially independent, epigenetic, axis of heredity and evolution. Only if this conjecture is empirically valid are the arguments presented compelling. I think that with a few exceptions most biologists and philosophers of biology today accept that there is more to heredity than genes, and that soft inheritance, including cellular epigenetic soft inheritance, exists. Not everyone agrees, however, that it is common, and this is an empirical question. In a recent survey conducted by Gal Raz and myself, we found over a hundred good cases of cellular epigenetic inheritance in bacteria, protists, fungi, plants and animals, and these seem to be only the tip of a great iceberg. It is clear that we need to learn more about epigenetic inheritance: the existing examples have to be more thoroughly studied, additional cases must be found, and the mechanisms underlying the induction and transmission of epigenetic variation need to be better understood. There is a need for comparative studies that will unravel the evolutionary origins of the epigenetic mechanisms, and point to generic processes of development that are important in the generations of phenotypic novelty. Moreover this has to be done not only at the cell level but also at higher levels of biological organization, including the behavioural and cultural levels. However, the conceptual inferences from our basic position, if valid, are fairly straight forward, although they are indeed far-reaching. These inferences are not generally recognized by evolutionary biologists, for reasons that, in my opinion, have more to do with intellectual and professional inertia than with logic.

Conceptually, the most difficult part of the epigenetic-developmental position is that related to cultural evolution in humans. It can be argued that it is misleading to use evolutionary reasoning in this case, that "cultural evolution" is a misnomer. I believe that evolutionary reasoning is useful for the human cultural case, because the basic concepts of exploration of a large set of possibilities, followed by the selective stabilization of some of those explored possibilities and their accumulation over generations, applies in the cultural case. However, I do recognize that symbol-based representation and communication occurs at so many levels of abstraction, in so many modalities, at so many levels of social interaction, and in the context of such varied social structures, that it is very difficult to think about it in the relatively simple evolutionary terms that can be applied to evolutionary systems based on cellular, heritable epigenetic variations or even to evolution based on the transmission of patterns of behaviour in animal societies. Symbol-based cultural systems are fundamentally different from animal traditions, although humans also have non-symbolic routes of information acquisition and transmission that are of great importance, and that interact during both early and late development with their symbolic system.

In spite of the complexity, it is possible to achieve important insights into the role of culture in evolution if we "black-box" the symbol-specific aspects of human culture and just treat their transmissible, trans-generational effects. For example, it is possible to understand the distribution of the tolerance to the milk protein lactose if we realize that the cultural practise of the domestication of cattle led to the use of fresh milk as an energy source, and that a stable culture of fresh milk usage had ramifications that led to the selection of genetic variants that enabled the humans who practiced this culture to have the advantage of lactose absorption. However, how the practices involved in the domestication of cattle were invented and transmitted, is not addressed by this type of scenario; the scenario would have been exactly the same (as long as the transmission was stable) if the practice was acquired and transmitted by much simpler, non-symbolic forms of social learning. "Black-boxing" thus neglects the generation of this novel practice, its establishment and its relation to the encompassing symbolic culture. We therefore miss some important aspects of the dynamics of its transmission that are related to its symbolic nature. I think that understanding human social and cultural history requires a consideration of the specificities of human cultural symbol-based evolution/development. DST (developmental system theory), a radical process-oriented approach to evolution developed by Susan Oyama, seems like the right framework for thinking about human culture. Symbol-based human evolution presents an enormous challenge, which demands a real interaction of evolutionary biologists with anthropologists, sociologists, psychologists and historians. The conceptual, evolutionary-historical framework that will eventually emerge from this effort may be very different from those that we employ when thinking about evolution at other levels.

5. With respect to present and future inquiry, how can the most important problems concerning evolutionary theory (or evolutionary aspects of your field) be identified and explored?

Classical problems of evolutionary biology pertain to processes and patterns of evolution. From the point of view that I have outlined, the questions are: does an epigenetic perspective make a difference to the way we understand adaptive evolution? to speciation and other macroevolutionary changes? to major evolutionary transitions? to rates and trends of evolution? What kind of research needs to be conducted to address these problems?

Marion Lamb and I have discussed the evolutionary implications of epigenetic inheritance in many of our publications, so I shall point to these implications and the research they require only very briefly.

Epigenetic variations can directly contribute to adaptation through the selection of heritable epigenetic variations. It is known that there is extensive epigenetic variation in natural populations, and lab experiments suggest that some of it can be beneficial. For example, inherited epigenetic differences in the color and morphology of flowers might be beneficial if new pollinators were introduced into their habitats; in bacteria, recent experiments suggest that inheritance of epigenetic variations in gene expression may underlie some forms of antibiotic resistance. If we think about the transmission of socially-learnt information in animal groups, a lot of adaptive animal traditions may be based on behavioral transmission. In order to explore the direct contribution of epigenetic variations to adaptive evolution within populations, it is necessary to conduct research on natural populations, estimate the extent of the epigenetic variation of interest, the stability of epigenetic variants, and the conditions of their induction or invention. Models evaluating epigenetic heritability and models of evolutionary dynamics based on epigenetic variations need to be constructed

The effect of epigenetic variation may be indirect, affecting the generation of variation in DNA: the rate of mutation, transposition and recombination is affected by the epigenetic state of the gene (its chromatin state). Another indirect effect is facilitating the selection of genetic variants, by accelerating the rate of genetic accommodation. Genetic accommodation is a process based on the plasticity of organisms, their ability to respond to new environmental stimuli in an adaptive manner. Through the selection of genes underlying the developmental capacity to respond to a new environmental stimulus in an adaptive way, a genetic constitution that makes the adaptation very easy to produce can be built up. Epigenetic inheritance that leads to the persistence of the induced response contributes to the directionality of selection and hence facilitates the generation of stable adaptations. Eytan Avital and I argued that these processes may be very important for the evolution of behavior in animals, and Daniel Dor and I developed the idea that it was an essential part of the evolution of the linguistic human capacity. The study of animal traditions, focusing on the origins of innovative practices and the conditions of their stable transmission in animals, is therefore of major importance.

Speciation may often be initiated by heritable epigenetic variations. For example, chromatin marks on the two sets of parental chromosomes may be incompatible as a result of different developmental histories of the parental populations, resulting in postzygotic reproductive isolation. Different behavioral traditions in groups of animals may also be a barrier to gene exchange and promote speciation. For example, the dialect of a song used by one group during courtship may sound unacceptable to members of another group, so mating between the members of the two groups will not occur.

Epigenetic control mechanism seem central to the generation of new species in another way, through polyploidization and hybridization, which are processes that play a major role in plant evolution. Following hybridization and genome duplication in plants, epigenetic control mechanisms are recruited, leading to genome-wide epigenetic variations, many of which are heritable. Moreover, extensive *genetic* variations may follow, targeted to the genomic regions that were marked and epigenetically modified by the epigenetic control mechanisms. The sequences modified are mainly transposable elements, ribosomal DNA, and other types of DNA repeats, which are genomic regions that play a role in the regulation of gene expression and in the control of chromosome-behavior. It is worth noting that similar mechanisms seem to be recruited in extreme ecological conditions, such as when organisms are starved, or their nutritional conditions are drastically altered, or there is a persistent climatic change like an increase in temperature, or as a consequence of hormonal stresses. The variations generated by these mechanisms are extensive and coordinated at the genomic level, and may often have large phenotypic effects. They may be an important source of innovation in adverse conditions, which are often the very conditions that are conducive to evolutionary change. If so, it may be possible to induce them under controlled conditions by exposing organisms to the relevant stressful environment or directly manipulating the generative epigenetic mechanisms. This means that it may be possible to study macro-evolution in the laboratory! The variations in chromosomal regulatory architecture that would be discovered are not expected to follow the rates and patterns of evolution shown by coding genes. This calls for research centered on chromosomal evolution, using the most advanced molecular tools.

Epigenetic inheritance must have been important in the major transitions of evolution identified by John Maynard Smith and Eörs Szathmáry. Marion Lamb and I have argued that epigenetic inheritance mechanisms were involved in the evolution of chromosomes, in the evolution of the compound eukaryotic cell, in the transitions to multicellularity, to social groups and to human groups using linguistic communication. Our perspective led to the suggestion that another transition – the transition to neural organisms – should be added to those enumerated by Maynard Smith and Szathmáry.

The implications of epigenetic inheritance both for adaptive evolution in populations and for macroevolution suggest that rates of evolution may be sometimes very high, and that there may be more directionality to evolutionary change than previously acknowledged. Rapid rates of evolution may in part be the outcome of rapid changes in the epigenome, not just intense selection, and directionality may be the result of the targeted nature of the epigenomic changes that occur under stressful conditions.

Epigenetic inheritance is clearly extending our present notions of heredity and of evolution. I believe that we are at an exploratory, exciting period in the history of biology, a time of changes and new syntheses.

10
Philip Kitcher

Professor of Philosophy
Columbia University; USA

1. Initial Impetus towards Evolutionary Theory

As a graduate student, my education in the philosophy of science followed a pattern that was standard in the late 1960s and early 1970s. The principal illustrations of general claims about the character of scientific explanation, confirmation, theories, and so forth came from the physical sciences. So, at the beginning of my teaching career, the examples I gave my students were drawn from physics and chemistry. Only when a public-spirited student pointed out to me that most of those enrolled in my class were studying biology did I begin to consider whether I might be providing a one-sided diet of examples. Fortunately, I found my way to David Hull's recently-published book, *Philosophy of Biological Science*, which not only offered me ideas for teaching my students, but also introduced me to fascinating material that led me to understand how central philosophical issues about science might be viewed very differently.

Prominent among the questions discussed in Hull's eye-opening book was a debate about the status of Darwin's evolutionary theory. As I learned more about evolution, both in its original (1859) formulation and in the version into which it had settled in the 1970s in the wake of the modern synthesis, it became evident to me that this was a major scientific achievement, one that philosophers simply could not dismiss, and that it did not fit easily into the philosophical categories I had been taught. My response was to conclude that those categories were inadequate, and that the general philosophy of science should be informed by Darwin's achievement.

So far, I still thought of myself as a general philosopher of science, oriented towards a major field of scientific work, neglected or dismissed by most of my colleagues. I hoped to re-orient what I then took to be the central questions about explanation, confirmation, and theories by using Darwinism as a test case. Of course, to do this thoroughly, I had to steep myself in evolutionary theory. I decided that an extended course in reading would not be enough, and, on my first sabbatical leave, in 1981-2, I went to Harvard's Museum of Comparative Zoology, where Stephen Jay Gould kindly gave me space in his laboratory. That year was full of discussions that changed my thinking in fundamental ways, and I shall forever be grateful to Steve, to Dick Lewontin, and to Ernst Mayr for their patience with me.

During that year, I became engaged in the debate about the credentials of evolutionary theory, and wrote my first contribution to the Creationism controversy. I also became aware of the many different ways in which biological claims were relevant to social issues. Steve and Dick had been active in combating the over-zealous sociobiological accounts of the springs of human nature, and, although I believed that they were broadly right, I thought that their critiques did not reach the heart of the issue. Further, I became aware of important theoretical controversies within evolutionary theory, debates about levels of selection, about adaptationism, about concepts of species. By the end of my stay, in the summer of 1982, my philosophical interest in biology had broadened enormously.

2. Distinctive Claims

I think of my work (perhaps wrongly – self-judgment is extremely fallible) as beginning with the limited project of using Darwinism as a test of philosophical ideas about theories and explanation. Most explicitly in my essay "Darwin's Achievement", I resist the thesis that a theory should be seen as an axiomatic deductive system, or even as a bundle of laws, in favor of the picture of it as a *toolkit*. In my more abstract philosophical treatments (which now strike me as suffering from an overly-fussy urge for a type of precision that is valued by a small coterie of philosophers but not important to anyone else) I propose that Darwin offered us schemata for answering instances of very general questions about life and its history. This idea can be made clear without the clutter of details: in response to questions about geographical distribution, for example, Darwin constructs narratives that show how current populations are the modified descendants of ancestral forms that underwent particular migrations; questions about the fit of organisms to their environment are addressed by elaborating narratives that trace the ways in which mutant forms arose in ancestral populations that lacked particular traits, and how those forms offered advantages in the struggle for reproduction. Instead of thinking in terms of general laws, or some "Principle of Natural Selection", we should see Darwin as setting an agenda of questions about the living world, and as providing directives for constructing answers. His successors have enlarged the agenda, articulated further the directives Darwin gave (for example, by showing how his general thoughts about the spread of advantageous traits must be understood in terms of the precise trajectories identifiable in population genetics), and, most evidently, in amassing a huge corpus of specific answers that exemplify his patterns.

This point may seem esoteric and inconsequential, but I think it has important implications. First, it defuses a certain family of objections that are often raised by critics of evolution, objections that center on the idea that Darwinism isn't a genuine scientific theory. In place of the thought that scientific theories are bundles of general laws, it sees science as an attempt to answer large families of questions, and theories as offering structured ways of doing that. Second, it opposes a strategy that pervades many attempts to draw Darwinian conclusions about human behavior. The sociobiologists of the 1970s and the evolutionary psychologists of today often tell their readers that there are general principles that support the provocative conclusions they draw – principles about mate choice or attitudes towards group members that hold across species, and that therefore must be admitted in our own case. Recognizing evolutionary theory as founded in the provision of particular explanations of specific traits in some definite species or range of species, we need to investigate whether the structure of the account given for the organisms that have been rigorously studied will apply to others. This means that evolutionary psychology cannot simply piggy-back off the careful work of those who study the behavior of non-human animals. Evolutionary psychologists must *themselves* do work that is as precise in its modeling, as painstaking in its accumulation of data, and as sensitive to alternative hypotheses as that undertaken in the large number of admirable evolutionary studies. Yet even in the most celebrated examples, the work of Leda Cosmides and John Tooby on detection of cheating, for instance, the modeling is pitifully crude in comparison to that found in mainstream evolutionary theorizing, the knowledge of environmental details is rudimentary and the underlying genetics simply absent.

My apparently limited methodological point has further implications for proposals about the "expansion" of evolutionary theory, and for supposed needs to reformulate its central ideas. Many critics of adaptationism, of evolutionary psychology, and of sociobiology have yearned for some simple way of identifying a general misunderstanding of the theory of evolution. By contrast, I view the troubles as local, results of the misapplication of the Darwinian toolkit, not deficiencies in the tools themselves. In various articles, I have argued against claims that there are definite levels at which selection acts, that there are fundamental conceptual flaws in seeking adaptations, that there are confusions in the standard evolutionary conception of the relation between organism and environment. The attitude I adopt is pragmatic. There are limited issues about levels of selection that arise when models of evolutionary processes are inadequate to capture causal details, but in almost all instances, I view us as having considerable choice as to the level at which we take selection to act.

Biologists know that certain kinds of specifications of fitness will not enable you to account for particular phenomena. To cite a well-known example, one cannot explain cases of heterozygote superiority in terms of context-independent assignments of fitness to individual alleles. On the other hand, by adopting a particular conception of the environment, one can give a causally adequate account in terms of allelic fitnesses that are relativized to the environment. Philosophers, and a few biologists, sometimes object that models of this sort somehow fail to capture the causal structure, that they do not pick out the trait *for which* selection is occurring. This seems to me a matter of mythology. Once one has recognized the causal processes that lead to a given effect, it is often possible to focus on different aspects of those processes, guided by pragmatic considerations of convenience. Some of Darwin's followers have been beguiled by what he explicitly saw as an analogy. The breeder selects by having a particular quality (or complex of qualities) in mind. If Nature "selects", it is not in virtue of such intentions. Natural selection results from the differential propensities of organisms to reproduce, or alternatively, of alleles to have favorable effects on their transmission – and one cannot insist that the processes through which selection occurs be described in one idiom or the other. That is to overwork Darwin's analogy.

Evolutionary biology rightly focuses on models of processes of selection that will explain the outcomes. The biology should not be cluttered with metaphysics. Nor should we conclude that Darwinism is incomplete until we have a multi-level theory of selection, or until we have recognized some organism-environment complex as the true object of the selective process. In my judgment, the tools of contemporary evolutionary theory are adequate to a large range of questions, and, despite fashionable charges, not demonstrably inadequate to cases that motivate the call for major reformulation. Instead, biologists and theorists should be alert for ways in which current tools are misapplied.

My approach to evolutionary theory thus sits between those of two groups that go in opposite directions. Sociobiologists and evolutionary psychologists claim that orthodox evolutionary theory is committed to sweeping conclusions about human nature. Critics and reformers seem to accept that supposed commitment, and reply that the theory needs to be purged of its adaptationism, needs to be expanded to allow for multi-level selection, needs to be reformulated to include the whole developmental system, or needs some similarly large reform. I deny the commitment. Evolutionary theory consists of a set of tools, and the tools are good ones. Maybe our toolkit should be enlarged, but I find no compulsion to expand it in the reasons the would-be reformers offer. I agree with their opposition to the attention-grabbing claims made by those who "dare" to apply evolutionary insights to our own species, but I offer a different diagnosis: these are abuses of the tools we have inherited from Darwin.

This pragmatism also makes me suspicious of what I take to be over-ambitious attempts to apply Darwinism in a direct way to philosophical questions. The difficulties of evolutionary accounts of morality are well-known, apparent in the dismal track record of ventures in "biological ethics" from the nineteenth century to the present. Those difficulties stem ultimately from the failure to take seriously details of human evolution. In similar fashion, the currently fashionable attempts to develop evolutionary epistemologies, the fascination with "cultural evolution" and "memetics", are most successful where they attend to details. Writers like Robert Boyd, Peter Richerson, L.L. Cavalli-Sforza, and Marc Feldman have shown how complex any serious approach to cultural evolution will have to be.

Pragmatism is, however, entirely compatible with a more limited view of the importance of Darwin for philosophy. Behind the prominent proposals about descent with modification and the action of natural selection stands a more basic Darwinian thesis, one shared with many other nineteenth century thinkers. The thesis of *Historicism* is that history matters to our understanding, that we can only comprehend certain phenomena if we are aware of the processes out of which they have emerged. Darwin made a brilliant case for Historicism with respect to living things. He suggested that we could not explain the distributions of organisms, the similarities and differences among their traits, the character of the fossil record, and the fit between organism and environment, without attention to the processes that have given rise to the organisms on which we focus. In my view, philosophy should make a similar commitment to Historicism. We cannot understand large features of our current practices – in Religion, in Science, in Mathematics, and in Ethics – without fathoming the processes that have brought those practices into their present forms.

Inspired in this way by Darwin, I advance a general philosophical position, *Pragmatic Naturalism*, that insists on viewing the questions philosophers discuss in the light of history. Just as Darwin elaborated as much of the history of life as he could, I think we should do the same with respect to scientific knowledge, mathematical knowledge, religion, and ethics.

3. Implications for ethics and social thought

Given what I have already said, it should not be surprising that I take the Darwinian account of the evolution of living things to have no direct implications for either ethics or social thought. Enthusiastic opponents of religion who take Darwin to pronounce the doom of faith are mistaken. Equally, those who attempt to derive ethical precepts from premises established in evolutionary biology commit fallacies.

Nevertheless, if you take philosophical inspiration from Darwin in the way I have suggested, some conclusions about religion and ethics are suggested. The issues here are complicated, but I hope that a simplified account can make the position I defend plausible. I shall begin with the topic of religion, and proceed to the (harder) case of ethics.

Some forms of religious belief are incompatible with the acceptance of scientific findings. Anyone who supposes that every sentence of the Bible must be taken literally is going to be in trouble with pre-Darwinian discoveries about the age of the Earth and with the early nineteenth century consensus that different communities of organisms have lived on our planet at different historical stages. A person with these beliefs will also find geological dating problematic — and, indeed, may wonder how to reconcile our understanding of the motions of the heavenly bodies with Biblical passages. Evolution is only part of a corpus of scientific ideas, each of which undermines scriptural literalism.

The specific thought that life has evolved over billions of years, and has been largely propelled by natural selection is also antithetical to ideas of divine providence. Why should it have taken the Creator so long to get to the point? Why did He choose to bring about His plan by processes that are so wasteful, and that involve, once sentient organisms are on the scene, so much suffering? Evolutionary theory poses a sharp challenge, one that believers in Divine Providence have to take seriously. Yet this is to intensify difficulties that have been familiar for centuries before Darwin wrote, and that were discussed almost at the dawn of Christianity. The most successful answer to them is to confess human ignorance, and that answer is available in the wake of the Darwinian exacerbation of the trouble.

Far more serious problems for religion emerge not from the evolutionary view of life but from the application of Darwin's most fundamental attitude to the case of religion. Historicism about religion invites us to consider how the religious practices of people around the world have come to be as they are. The first point to recognize is an extreme diversity of doctrines. There are religions with a single divinity and religions with many, religions that invoke the abiding presence of ancestors and spirits, religions based on impersonal supernatural entities, forces with which the faithful should align themselves. That diversity is coupled to an extraordinary similarity in the ways in which religious practices have been transmitted from the past to the present. The world's religions typically suppose that there was a time in the past when some privileged people were given access to the supernatural world. From the first revelations, the religious truths have been transmitted, across the generations, to the devout of today. Even if further revelations arise, as they are often claimed to do, it is notable that they are interpreted, in each tradition, in terms of the orthodoxy that has been passed down.

What follows? First, from an epistemic viewpoint, all religions are on a par. The processes that undergird Christian faith are no better, and no worse, than those which support the beliefs of the Nuer, or the Inuit, or the Crow indians, or aboriginal Australians. Second, the doctrines are all incompatible with one another. Not all of them can be true, and there is no significant core of beliefs about the supernatural, shared by all. That should already inspire skepticism about any particular claim – even the bare thesis that there is a supernatural "something". After all, if the convinced Christian had been brought up among the aboriginal Australians, he would have as firm a commitment to the Dreamtime as he actually does to the Resurrection – and that commitment would have been produced by similar processes, processes of epistemically identical value.

A closer look at the actual processes through which religious doctrines are transmitted, and how they are modified, deepens the problem. A host of allied sciences has scrutinized the formation of canonical texts, the recruitment of new members to religions, the intertwining of religious doctrines with social and political questions, the circumstances under which people claim religious experiences. This body of research shows with enormous clarity that the processes through which the religions of the world have descended to the present have much to do with human psychological and social needs, but that they have very little to do with the discovery and transmission of truth. The success of religions depends on their being good at responding to felt human needs. From the investigations I have mentioned, it becomes evident that religions do not have to be true to be good.

Historicism about Religion supports a thesis that lies between agnosticism and atheism. The agnostic part derives from the facts that science is incomplete and that we are fallible: one cannot rule out the possibility that some future amendment of our current picture of the world will disclose some entity reasonably described as supernatural. The atheist part is simply a commentary on the actual views of the supernatural we have: *all* of them merit our disbelief and rejection. Even if we envisage the possibility that future science might vindicate the idea of the supernatural, we should be very confident that the primitive myths that have come down to us are entirely inadequate representations of it. So there's a bare possibility that some claims about the supernatural will turn out to be true, but overwhelming evidence to think that all those claims that human beings have so far made are completely false.

Nothing I have said has any implications for those forms of religion that do not require a literal belief in anything supernatural. If you think that the important religious stance is some other attitude rather than belief (so that there is no substantive doctrine, but something more like a commitment to valuing particular things) or that the scriptures are apt metaphors for the human condition, then your religion is unscathed by my conclusions. Many religious people suppose that "liberal" forms don't really count as religion, but I am happy to allow non-literalists to use their preferred terminology. More importantly, I think secular humanists should welcome liberal denominations as allies, for they show the possibility of ways of life that abandon superstition without jettisoning practices that have been central to every human society about which we know.

Many critics of religion, including those who most want to press an evolutionary case, think that the abandonment of religious myths will be a salutary development. They are partly right: to give up false beliefs is usually a good thing. They can (and do) frequently go further, arguing that these false beliefs have caused significant harm, particularly in distorting the lives of people who have to live according to alleged divine commandments and in the hatred and violence that doctrinal differences have brought in their train. So far, so good, but militant secularists often fail to see the ways in which religions have *responded* to human needs, and how the removal of religious *practice* would create a vacuum, into which old myths would be likely to intrude. Ironically, given the Darwinian enthusiasms of some of these critics, they fail to wonder if a form of social life that is found in so many cultural environments and that has spread so successfully might have something important to recommend it.

I suggested that religions don't have to be true to be good. I also concluded that they aren't true. It doesn't follow that they aren't good, where goodness is understood in satisfying human needs. The challenge for secularists is to identify the genuine functions that religions have served, and to provide thoroughly enlightened substitutes for them. Militant atheism often offers *secularism*, but what we really ought to aim for is secular *human*ism. In my judgment, the real conclusion to be drawn from Historicism about Religion, or, if you like, a Darwinian reconceptualization of this area of human life, is the need for positive responses to deep features of the human condition that do not lapse back into myth and superstition.

I turn now to the harder case of ethics. Evolutionary attempts to understand ethics range from the crudest suggestions of sociobiology ("We shouldn't sleep with our siblings because it is selectively disadvantageous!") to the more sophisticated thought that our evolutionary past has set up in us tendencies (possibly altruistic tendencies) whose (proper? normal?) expression is in ethical conduct. I begin from Historicism about Ethics. Where do our ethical practices and our explicit ethical codes come from? In agreement with the suggestions of some (but by no means all) primatologists, I suppose that we, our closest evolutionary relatives, and our common ancestors, shared dispositions to *psychological* altruism (that is: to tendencies to respond to the predicaments of others, to form intentions to remedy those predicaments without concern for selfish advantage, and to act on those intentions). Although such tendencies made it possible for our ancestors to live in small societies mixed by age and sex, I believe that they were limited. Despite a propensity to respond to allies in *some* contexts, there were recurrent occasions on which our ancestors let one another down, and in consequence they needed to engage in time-consuming gestures of peace-making and mutual reassurance. If all we had to go on were our altruistic dispositions, our societies would still be small, our social lives tense and fragile, and our opportunities for cooperation quite restricted.

A great step in human evolution, coeval in my judgment with the acquisition of full language, came with an ability to foresee consequences of our own socially-disruptive actions, and to inhibit ourselves. I envisage small bands of early humans discussing their behavior with one another and formulating guidelines for action. Out of their deliberations came the first ethical codes, and, for at least 50,000 years, human societies (at first very small, in the last 5000 years substantially larger) have been experimenting with those codes. Some rules have emerged in almost all societies, and remain stable after their emergence. On other issues, there are unresolved differences.

I can't expect to make this picture of the history of our ethical practice plausible (that would require *far* more space than I have!), but if something akin to it is correct, there are implications for how we think about ethics, and perhaps suggestions for how we go on with our ethical practice. The first point to note is that ethical achievements do not consist in discovering some body of independent truth. Ethics is a human invention, a form of social technology in which groups of people react to their felt difficulties. As we make advances by addressing our current difficulties, new problems are generated, new concepts of ourselves, and new forms of society emerge. The process is never finished. Moreover, in its early stages, the fashioning of the ethical life was a matter for joint deliberation. There was no thought that a single individual

– a religious leader or some extraordinarily insightful philosopher – could justifiably pronounce on ethical truth. If we are to talk of ethical truth at all, the best sense we can give is of rules (and associated evaluations of action) that emerge in a wide variety of societies and that remain stable thereafter. On this account, it is possible that only some ethical statements are either true or false, and that others are indeterminate.

Perhaps this awareness of our ethical past can help us more reliably and more self-consciously to evolve our ethical practice. If, as I believe, the first 40,000 years of ethical experimentation proceeded on the basis of discussions among people who recognized themselves as having to interact with one another, deliberations in which all participants were taken to be essential and thus equal, then we might inquire how we should renew the practice of ethics in a world in which the lives of many people are bound together. Ethics can no longer be a matter of relying on the declarations of alleged authorities. Rather, the ideas advanced in religious texts and philosophical treatises should be viewed as proposals, thoughts to consider in our joint discussions. Living in a very different social world, we should, I suggest, think of all members of our species as parties to that deliberation. Each of us has equal status in it, and, in a world that offers opportunities for us not only to meet our basic needs but also to articulate and develop our own projects for our lives, the pertinent standard of equality might be characterized as equal opportunity for a self-determined, worthwhile life.

The tentative proposals of the last paragraph should not be regarded as implications of the Historicist picture, let alone implications of Darwinian discoveries about the evolution of living things. Instead, they arise from the sense that ethics can *only* be a matter of proposal and joint discussion, as suggestions about the ways in which the discussion might proceed, and the ideal at which it might aim. Historicism about Ethics simply offers a perspective from which our ethical practice might be viewed differently, and, given that view, we might pursue it in new ways.

4. Interesting Criticisms

Before I confront what I think of as the most important objection to my views, I would like to address what I think of as an extremely dull and misguided critique that has given rise to far too many philosophical articles. The essay, "The Return of the Gene", co-authored with Kim Sterelny, has inspired many responses, all of which claim to show that genic models just cannot be adequate to certain types of evolutionary processes. Almost all of these miss the major point of our article, to wit that the same causal story can be told in different ways. The critics are inspired ultimately by a real insight of Elliott Sober's, who pointed out that there can be selection *of* Xs without there being selection *for* the property of being X. Sober illustrated the point by describing a "selection toy", a cylinder divided by four different planes, each of which contains holes; the size of the holes decreases as one goes from top to bottom; the result is that balls of different sizes will fall through and be stopped at different levels; the smallest balls will fall all the way to the bottom. Supposing that the smallest balls are pink, there is selection *of* the pink balls; but there isn't selection *for* being pink; there is, instead, selection *for* the property of being smaller than a certain size.

Adverting to the selection toy, one can see that a model might show that organisms with a specific trait are selected without showing that there is selection *for* that trait. A claim often made in discussions of the "units of selection" is that Richard Dawkins, Sterelny and I have confused selection *of* with selection *for*, that the models we view as possible "keep the books" but don't present the "causal structure". But this is simply misunderstanding. The point isn't that *every* model of a particular sort that gives the right gene frequencies is adequate, but that *some* are. Extend the account of the selection toy to make it more "lifelike". Imagine that the balls *grow* from the same initially tiny size. When they are first put into the (inverted) cylinder, they sit at the bottom. There are four types that grow at different rates through the same period. The A-type is the slowest growing, and thus is the smallest when the growth period ends. The A-type balls are also pink. When the toy is set up and the balls fall through, the balls that end up at the bottom have three properties: they are pink, they are the smallest, and they are A-type. As Sober said, we have selection *of* the pink balls, but no selection *for* being pink. Do we have selection *for* being small, or selection *for* being A-type? That is one question too many. The causal structure can be adequately captured either way – for the obvious reason that there is a causal connection: the smallest balls are smallest *because* they are A-type. By the same token, accounts that identify the allelic combinations that cause (in the pertinent environments) phenotypic traits we would regard as targets of selection are as causally adequate as those that focus on the phenotypes. We are identifying the same causal process with two different emphases. Since Nature isn't a selecting agent, with fine-grained intentions, there is no fact of the matter about what "She had in mind".

I have dealt with this "boring" objection, because it is so common, that not reacting to it might appear to be evasion on my part. The serious objection – which may, I suspect, lie behind the propensity of my critics to press the boring misunderstanding – is that the pragmatism I espouse is too devoted to evolutionary theory as the majority of biologists worldwide conceive it and practice it. My pragmatism can be seen as a conservative opposition to important new innovations. Champions of sociobiology and evolutionary psychology view me as using the comparison with standard modes of evolutionary theorizing to strangle promising nascent sciences. They might admit that they cannot yet do work on human behavior that is as fully developed and as rigorous as that offered by those who look at the evolution of a well-studied trait in a tractable species (coloration in guppies, say). If their attempts to fathom human psychology are to make progress, then they will have to be allowed time to elaborate their currently embryonic insights. By the same token, many of those who have campaigned eloquently against proposed evolutionary analyses of human behavior look for something that will bring what they see as politically noxious science to a halt. They claim that there is something fundamentally amiss with contemporary evolutionary theory, and propose to fix it by insisting on a proper understanding of the "units of selection", or a reformulation of evolutionary theory to embrace "multi-level selection", or a more thorough integration of development into evolutionary theory, or a refined understanding of the notion of environment. My approach leaves evolutionary theory as commonly practiced, and thus allows for the perpetuation of alleged "evolutionary sciences" of human beings.

My general approach to the philosophy of science recognizes the importance of diversity of opinion within the scientific community, so I am troubled by the thought that my position might exclude valuable options. I would be less concerned to hold evolutionary studies of human psychology and behavior to high standards, if we lived in a world in which the findings of such studies were soberly presented as preliminary, in which the difficulties of devising adequate evolutionary models of human psychology and behavior were clearly understood and acknowledged, in which "results" were not trumpeted to the general public as exciting new discoveries, firmly established, and in which they were not so frequently directed towards reinforcing harmful stereotypes about particular groups of human beings. Since the conditions of our world are so different, it seems to me important to emphasize the shortcomings of the supposedly illuminating "new perspectives". I would also be more sympathetic towards proposals for extending and reformulating evolutionary theory if they showed some signs of delivering observational and experimental findings beyond the reach of the prevailing orthodoxy. In general, I worry that the proposals for reform are generally grounded in metaphysical pictures, rather than stemming from the pressure of unsolved scientific problems. The metaphysics seems to me to be muddled, and the pictures thus have little impact on the actual work that evolutionary biologists do.

There are obvious dangers, however, that my pragmatic response may be intolerant of novelty and overly complacent. Pragmatism should be scrutinized as carefully as it scrutinizes the ventures about which it is skeptical. I hope it can live up to the high standards it sets for others.

5. Future Inquiry

Despite the enormous body of evidence that supports it, many people remain skeptical about Darwinian evolutionary theory. One task for philosophers, as well as evolutionary biologists, is to continue to work for broader public understanding of the evidence and broader awareness of the fallacies and mischaracterizations that pervade the works of contemporary creationists (including those who advocate what they call "Intelligent Design"). This is the "Lord's work" – and it is never finished.

Philosophy has a special role to play in fostering greater understanding of Darwinism for the very obvious reason that opposition so often turns on the supposition that evolutionary theory is deeply inimical to things people hold dear. The issues here are complex, and public understanding is aided neither by militant claims that Darwin makes religion obsolete nor by confused attempts to delineate two separate spheres, in one of which supernaturalist religion can find asylum. As I see the challenge, philosophers need to return to the deepest problems of philosophy (centered on questions about how to make sense of our world and how to live in it). Guided by the evolutionary theory we inherit from Darwin, among other achievements of our inquiries, we should try to develop a fully humane secular humanism, one with which people can contentedly live. That is an enormously complicated and difficult task, and the enduring controversies about the status of evolution are one – very minor – symptom of the fact that we are very far from completing it.

11
U. Kutschera

Professor of Plant Physiology and Evolutionary Biology

Institute of Biology, University of Kassel, Germany

Evolutionary Biology: A System of Theories

1. Why were you initially drawn to discussions and research on evolution (or evolutionary aspects of your field)?

Evolution, i.e. descent with modification, is not "just a theory", as creationists and proponents of the intelligent design (ID) movement claim, but an ongoing process that took place in the past and can be reconstructed. In some cases, evolutionary changes can even be analyzed experimentally, for instance, with populations of bacteria cultivated *in vitro*. I would like to add the following remarks on this important issue.

I. After more than 150 years of research, biologists and geologists distinguish between the *fact* of evolution (in Darwin's words of 1859: "descent with modification") and the *mechanism(s)* that bring about the documented adaptations and diversifications in populations of evolving pro- and eukaryotic micro- and macro-organisms (according to C. Darwin and A. R.Wallace 1858/1859: "natural and sexual selection" in groups of animals and plants). This important distinction is largely lacking in many discussions and popular articles on topics such as "Darwin and evolution" (Gould 2002, Mayr 1982, 2001).

II. In the natural sciences, theories explain sets of well-established facts. Hence, the first and most important task should be to teach pupils and the general public why *macroevolution* – the principal target of the Young-Earth- and ID-creationists – is a fact and not "only a theory". Here, cell biology (the principle of endosymbiosis) and paleobiology (transitional forms in the fossil record) are the disciplines of central importance (Kutschera 2008 a, 2009 a, b; Kutschera and Niklas 2004; Gregory 2008).

III. Evolutionary biology, which evolved from Darwin's famous three books published in 1859, 1868 and 1871, is a system of theories that aims to explain the origin and diversification of life on Earth. This interdisciplinary branch of the biological sciences encompasses both chemical and organismic evolution. It is obvious that evolutionary biology can not be reduced to molecular phylogenetics and an associated narrow view on the genotype of extant organisms. In evolutionary research, virtually all major branches of the life sciences are of equal importance: taxonomic studies at the organismic and molecular level (comparative morphological and DNA-sequence analyses), cytological investigations (endosymbiosis and cell evolution), comparative embryological, physiological and ethological studies (evolutionary developmental and behavioural biology), insights from the Earth sciences (plate tectonics, geochronology, paleobiology, mass extinctions), microbiological investigations (experimental evolution of bacteria and ribozymes *in vitro*), computer simulations (evolution of digital organisms, i.e., the new *in silico* biology) etc. (see Kutschera and Niklas 2004, 2005, 2008 for a more complete list). In other words, both the organismic and molecular aspects of evolutionary biology should be represented and integrated. In my view, a sound knowledge of the diversity of living organisms (classical taxonomy) is the key to an understanding of evolution. The shortsighted "molecularization" of biology has resulted in the almost complete elimination of biodiversity research at the universities throughout Europe and the United States: The consequence of rising illiteracy in taxonomy among many biologists and science teachers is good news for the anti-evolution movement around the world.

After these general statements I want to provide a short answer to question Nr. 1: My research in the area of evolutionary biology is purely curiosity-driven. As a 21-year-old biology student at the University of Freiburg (Germany) I discovered by chance that in populations of freshwater leeches of the genus *Erpobdella* , maintained in aquaria, every individual maximizes its fitness via the destruction of the cocoons of its breeding conspecifics (intraspecific cocoon cannibalism). This discovery led to my first scientific publications on the taxonomy and reproductive behaviour of aquatic annelids that later gave rise to a novel theory on the evolution of parental care, inclusive of the feeding of young, in freshwater leeches (Kutschera and Wirtz 1986, 2001).

2. What does your work reveal about biological evolution (or evolutionary aspects of your field) that other academics, citizens, philosophers or biologists typically fail to appreciate?

I started my career as a zoologist/evolutionary biologist three decades ago when I observed, documented and analyzed the unique behaviour of aquatic leeches (see above). Moreover, I discovered and described several new species. For instance, the "Golden Gate Leech" *Helobdella californica* Kutschera 1988, a taxon I found in 1986 in ponds of the Golden Gate Park in San Francisco, California, is an accepted annelid species of the United States and listed in the ITIS-report. These field studies, which I have continued to the present day, were later supplemented by the use of molecular techniques, including DNA-barcoding and reconstruction of phylogenetic trees based on mitochondrial gene sequences. Hence, in my work in the area of evolutionary ethology, which is based on (but not restricted to) leeches, I combine classical biology with molecular techniques. I should mention that I earned my master's degree in zoology/evolutionary biology and my Ph. D. (Dr. rer. nat.) in plant physiology. In this second area of research we not only analyzed the biophysical and molecular basis of the action of plant hormones, but we also studied the development of chloroplasts and mitochondria in juvenile seedlings of higher plants. This experimental work led to my ongoing interest in the concept of symbiogenesis, which is also known as the theory of endosymbiosis (Kutschera 2009 a, b).

Within the context of our studies of plant growth and development, which led to a modified version of the "green lineage concept" for the evolution of land plants, we discovered epiphytic bacteria of the genus *Methylobacterium* that consume methanol released by the plant. Moreover, these microbes synthesize and secrete phytohormones. In bryophytes, which are "living fossil plants", methylobacteria form a symbiosis with their corresponding eukaryotic host organism (Kutschera 2007 a).

This broad interest in evolutionary biology, from ethology/systematics of animals (leeches, fishes) via the structure, function and inheritance of organelles of eukaryotic cells (endosymbiosis) down to prokaryotic symbiotic bacteria that colonize land plants, led to several of my publications wherein we extended the synthetic theory of biological evolution (Gould 2002, Mayr 1982, 2001, Haffer 2007) and proposed an "expanded synthesis" (Kutschera and Niklas 2004, 2005, 2008).

I do not claim that other biologists would not have performed empirical studies of the same scope, but this broad experimental/analytical approach to evolutionary phenomena was the key to my success as a scientist working in Germany (University of Kassel) and, since 2007, in the United States (Carnegie Institution for Science, Stanford University, California). Likewise, my recent summary of the "evolution of Darwinism" (Kutschera 2008 a, b) was largely based on these empirical studies in different areas of the biological sciences. Colleagues who are specialists in one narrow field sometimes fail to recognize the interdisciplinary character of modern evolutionary biology, which is a system of theories from the geological and life sciences.

3. What, if any, practical and/or social-political and/or moral obligations follow from your work on evolution?

My work on different aspects of micro- and macroevolution led to papers describing the expansion of the synthetic theory of the 1950s and the definition of evolutionary biology – a term coined in 1942 by Julian Huxley – as a system of interrelated theories (Kutschera and Niklas 2004, Kutschera 2008 a, b). Moreover, I have pointed out repeatedly in my books, scientific articles and published interviews that biological evolution is a fact of nature. Creationists and proponents of the ID-dogma have modified this sentence as follows: "U. Kutschera claims that evolutionary *theory* is a fact – he confuses theories with facts". I have never made such a statement. Based on this perversion of my sentence, the German "Anti-Darwinists" argue that I would promote a dogmatic form of "evolutionism", which they compare with a political/religious ideology. This claim is not justified, but persists in the anti-evolution literature of Europe.

Two practical/socio-political consequences follow from my research in evolutionary biology.

First, my studies on the behaviour and biodiversity of annelids led to the discovery of new leech species, and the documentation of some of those taxa, described in the literature of the 19th century, still exist today. For instance, it is known that the medicinal leech (*Hirudo medicinalis*) largely disappeared from European ponds during the 20th century, due to habitat destruction. This economically important species has been confused with the related taxon *Hirudo verbana* from Italy and Turkey (Kutschera 2007 b). Like *H. medicinalis*, *H. verbana* is endangered, but not listed in nature conservation conventions. Moreover, the continuous import of these blood-sucking annelids to Germany means that an alien species is introduced into our fauna (leeches frequently escape into the wild or are thrown into ponds after use by the practitioner or patient). In addition, our recent search for the elusive European land leech (*Xerobdella lecomtei*) led to the discovery that a local population of this rare annelid from Austria became largely extinct, due to climate warming over the past four decades. When we failed to find adult *X. lecomtei*-individuals in the forests around Graz, Austria, we observed that the soil, wherein several of these rare animals have been collected about 40 years ago, was very dry. An analysis of climatological data provided the answer to this enigma: The air temperature has increased by + 3,0 °C over the past decades. Hence, climate warming was responsible for our failure to find adult individuals. Only one juvenile European land leech was discovered by chance, observed in the laboratory and analyzed (the dead worm was finally used to extract nucleic acids in order to sequence part of the mitochondrial genome for DNA-barcoding). Hence, our "Lonesome George of the annelids" lives on as GenBank No. EF 125040 (Kutschera et al. 2007).

Second, my theoretical papers and books on evolution vs. creationism led to the cessation of a continuous production of German Anti-evolution films (Videos, DVDs) produced by a company in Berlin. As the chairman of the German Association of Evolutionary Biologists, I could exert significant pressure on the "ID-creationists" in our country, although these religiously indoctrinated people rapidly adapt and evolve new strategies to promote their irrational claims.

4. What do you see as the most interesting criticism against your position in the biological or philosophical discussion of evolution?

It has always puzzled me that even educated persons with no special expertise in evolutionary biology (sociologists, physicians etc.) often fail to understand the distinction between scientific facts and religious myths. They stick to certain political/religious ideologies and hence are insensitive to all rational arguments proving that macroevolution occurred and is an ongoing process. In my new book, published in the "Year of Darwin" (Kutschera 2009 a), I describe another "proof for the fact of macroevolution", but I doubt whether this train of thought will change the opinion of the general reader. It is interesting to note that religious indoctrination of children at home, in Kindergardens and in schools may be one major cause for the fact that about 38% of adult Germans are creationists or adherents of the ID-dogma (Kutschera 2008 c). Hence, anti-evolutionism in all its versions must be viewed as a "faith-system" that is taken up by the youthful brains. Science education, which starts relatively late in public schools, can hardly overcome the religious indoctrination of our children.

The most provoking criticism against my position that biology should not be mixed up with any religious or pseudo-scientific belief system is the irrational argument that "scientists will never know everything; therefore, Biblical myths may be true and become part of evolutionary biology". This statement of some German creationists has been refuted many times and therefore should no longer be taken seriously.

5. With respect to present and future inquiry, how can the most important problems concerning evolutionary theory (or evolutionary aspects of your field) be identified and explored?

Evolutionary biology deals with the questions as to the origin of the first forms of life (chemical evolution) and the subsequent phylogenetic development of these earliest, ca. 3500 billion year old prokaryotic cells. In other words, evolutionary biologists ask the questions "how did the first cells emerge from the primordial hot soup of the young Earth" and "how did these proto-cells give rise to the extant (and extinct) bacteria, amobae, algae, fungi, animals and plants?"

The "origin-of-life question" has not yet been solved, although recently published, modified Miller-Urey-experiments yielded all of the 20 amino acids found in living organisms (Johnson et al. 2008). Moreover, these authors argue that their "Volcanic apparatus experiment" indicates that, on the young Earth, amino acids that formed in volcanic island systems may have accumulated in tidal areas where they could be polymerized to peptides by a simple volcanic gas. However, the question how these molecules developed into self-replication organic compounds is unanswered. Nevertheless, the principle of molecular self-assembly in aqueous solutions may be the key to this problem.

As to the further phylogenetic development of these earliest prokaryotic cells, symbiogenesis (primary endosymbiosis) was a key macroevolutionary event (Kutschera 2008 a, b; 2009 a, b) , but more work is required to further elucidate and reconstruct these historic processes.

References

Darwin, C. (1859) On the Origin of Species by Means of Natural Selection, or the Preservation of Favoured Races in the Struggle for Life. John Murray, London (6. Ed., 1872).

Gould, S. J. (2002) The Structure of Evolutionary Theory. Harvard University Press, Cambridge, Massachusetts.

Gregory, T.R. (2008) Evolution as fact, theory and path. Evo. Edu. Outreach 1, 46 – 52.

Haffer, J. (2007) Ornithology, Evolution, and Philosophy. The Life and Science of Ernst Mayr 1904 – 2005. Springer-Verlag Berlin, Heidelberg.

Johnson, A. B., Cleaves, H. J., Dworkin, J. P., Glavin, D. P., Lazcano, A., Bada, J. L. (2008) The Miller volcanic spark discharge experiment. Science 322, 404.

Kutschera, U. (2007 a) Plant-associated methylobacteria as coevolved phytosymbionts: a hypothesis. Plant Signal. Behav. 2, 74 – 78.

Kutschera, U. (2007 b) Leeches underline the need for linnaean taxonomy. Nature 447, 775.

Kutschera, U., Pfeiffer, I., Ebermann, E. (2007) The European land leech: biology and DNA-based taxonomy of a rare species that is threatened by climate warming. Naturwissenschaften 94, 967 – 974.

Kutschera, U. (2008 a) Evolutionsbiologie. 3. Auflage. Verlag Eugen Ulmer, Stuttgart.

Kutschera, U. (2008 b) From Darwinism to evolutionary biology. Science 321, 1157 – 1158.

Kutschera, U. (2008 c) Creationism in Germany and its possible cause. Evo. Edu. Outreach 1: 84 – 86.

Kutschera, U. (2009 a) Tatsache Evolution. Was Darwin nicht wissen konnte. Deutscher Taschenbuch Verlag, München.

Kutschera, U. (2009 b) Symbiogenesis, natural selection, and the dynamic Earth. Theory BioSci. 128: 191 – 203. (in press)

Kutschera, U., Niklas, K. J. (2004) The modern theory of biological evolution: an expanded synthesis. Naturwissenschaften 91, 255 – 276.

Kutschera, U., Niklas, K. J. (2005) Endosymbiosis, cell evolution, and speciation. Theory Biosci. 124, 1 – 24.

Kutschera, U., Niklas, K. J. (2008) Macroevolution via secondary endosymbiosis: a Neo-Goldschmidtian view of unicellular hopeful monsters and Darwin's primordial intermediate form. Theory Biosci. 127, 277 – 289.

Kutschera, U., Wirtz, P. (1986) Reproductive behaviour and parental care of *Helobdella striata* (Hirudinea: Glossiphoniidae): a leech that feeds its young. Ethology 72, 132 – 142.

Kutschera, U., Wirtz, P. (2001) The evolution of parental care in freshwater leeches. Theory Biosci. 120, 115 – 137.

Mayr, E. (1982) The Growth of Biological Thought. Diversity, Evolution, and Inheritance. Harvard University Press, Cambridge, Massachusetts.

Mayr, E. (2001) What Evolution Is. Basic Books, New York.

12
Richard Levins

John Rock Professor of Population Science

Harvard School of Public Health and the Instituto de Ecologíay Sistemática, Havana, Cuba

---

1. Why were you initially drawn to discussions and research on evolution (or evolutionary aspects of your field)?

My grandfather was a self-educated socialist who believed that every worker should know at least evolution, cosmology, and history. His views of evolution were Lamarckian, and that primed me to be fascinated by Lysenko's critique of genetics that seemed so dynamic compared to the "bean bag genetics" we got in high school. At the same time I met Marxist dialectics through the works of Haldane, Bernal, Needham, and Oparin. I dreamed interactions, connectivity, contradiction, wholeness and developed an aesthetic of complexity. In college I tried some Lysenkoist experiments (graft hybridization of tomatoes, the effects of pollen mixtures in corn) that turned out inconclusive for technical reasons. Years later, while a farmer in Puerto Rico, I rejected Lysenko because he located the origin of new species in individual development where I saw it as a population-level process. I found that Waddington and Schmalhausen had a better view of the feedback between individual development and ecology with the population and community levels. I decided that perhaps I could contribute something to science, and applied for graduate school.

2. What does your work reveal about biological evolution (or evolutionary aspects of your field) that other academics, citizens, philosophers or biologists typically fail to appreciate?

Most of the main elements of my work are now part of the common sense of the field:

1. There is a overlap of evolutionary and ecological time so that natural selection has to be seen in the context of changing conditions.

2. Species adapt not only to specific features of the environment but also to its pattern of variation in space and time, to uncertainty, patchiness, etc. This made adaptive strategy a central object of research. Faced with opposing demands, a species may adapt to some compromise phenotype, adapt to one demand while ignoring the other, or adopt a mixed strategy. This was expressed through the fitness set.

3. Organisms select, transform, transduce, define and respond to their own environments. The organism and its environment evolve together.

4. The response of a species to changing conditions takes place within communities of species so that an environmental impact on any one member percolates through the network of interactions, is damped along some pathways, amplified along others, and even inverted. The responses of interacting species can be studied through the community matrix, signed digraphs (loop analysis), and time averaging.

5. This lead into an explicit investigation of the strategies of modelling, especially qualitative mathematics, to determine what we can get away with not knowing and still understand a system. From this perspective the task of mathematics is to educate the intuition so that the obscure becomes obvious and even trivial.

6. Evolution and population dynamics proceed at multiple levels that may conflict. The direct impact of the environment may push the phenotype in one direction while natural selection moves it in another. For instance, along a gradient from the coastal desert to the rainforest in Puerto Rico, Drosophila melanogaster individuals are about the same size. Higher coastal temperatures make them smaller, but the selective pressure of desiccation makes them genetically bigger. In other cases, selection and direct impact push in the same direction so that species differ in nature in the same direction as they differ under controlled conditions, but more so. The meta-population concept expresses the possible opposing processes at group and mendelian levels of selection.

7. Biogeography is the result of the interaction among colonization, extinction, and speciation. Robert MacArthur, E.O. Wilson and I had planned a division of labor in which they would ask "how many species are there on islands?" while I would ask, "How many islands does a species occupy?" It was intended that both approaches would converge in a continental biogeography. Although this program was thwarted by Robert's early death and disagreements with Ed over socio-biology, it remains a valid strategy.

3. What, if any, practical and/or social-political and/or moral obligations follow from your work on evolution?

a) Anti-reductionism. No one level is more "fundamental" than any other. If gene action may determine which molecules are synthesized, the organism's conditions can determine which genes are active, the ecology determines those conditions, and past history gives us the genes that are there. Further, society is the evolving context for social change and individual's conditions. Social arrangements are fluid, and it is a mistake to attribute current, even widespread, phenomena to some fixed "human nature". Once social conditions are seen as co-variable with ecological processes, we can ask the question: how do we build a society where it makes sense to be kind?

b) The state of our science, with its pattern of powerful insights in the small and irrationality at the level of the enterprise as a whole, is also a social product not dictated by nature. It has a dual nature as the unfolding of human understanding and as the evolving product of a knowledge industry much conditioned by its owners. The examination of the social context of scientific priorities and beliefs is a necessary part of making scientific decisions. Thus our group criticized the expectation that infectious disease had been defeated in principle. The idea of the epidemiological transition was a major error, caused by several kinds of narrowness: extrapolation from short span of history and limited geographic range, limitation to a single species, ignoring of evolution and ecology, and the passive acceptance of "development" as a unilinear process in which the "less developed" recapitulate the history of the more developed. The green revolution was another failure due to narrowness and a static acceptance of boundary conditions of a market economy (capitalism). It promoted an agricultural evolution from labor to capital intensive, small scale to large scale, heterogeneity to homogeneity, dependence on nature to domination of nature, and ignored issues of land reform in favour of market fundamentalism. But it was caught by surprise: pesticides created pests, mechanization degraded the soil, plant breeding reduced genetic resources, technology increased class inequality in the countryside. Our alternative approach toward an ecological agriculture promoted knowledge-intensive agriculture with a mosaic of land uses and the nudging of natural systems of soil fertility and pest management, all in the context of a land reform. This approach to agriculture contributed to the Cuban adoption of ecological and organic farming.

c) Waddingtonian homeorhesis makes the study of vulnerability central to understanding health. From this perspective, the observed variance in a population is a result of the extent to which people's physiology is perturbed compared to the restoring forces (homeostasis) available. In poor, marginalized communities there is a more rapid erosion of homeostasis and also increased perturbation. In these conditions, small differences in individual circumstances have big effects. The question is therefore not why some poor people do well and others badly (the conservative, individualistic question), but what are the conditions that increase vulnerability.

4. What do you see as the most interesting criticism against your position in the biological or philosophical discussion of evolution?

a. My early work was hyper-selectionist. It focused on the question, what would the optimum condition be although I claimed only that species in nature would be expected to differ in the same direction as their optima. But I failed to consider processes such as: hitch-hiker genes carried along by selection at linked loci; epistatic constraints; slow gene frequency change near an adaptive minimum; random fixation of genes; multiple equilibria and selection toward extinction within communities.

b. It is often asserted that modern computation makes it unnecessary to simplify or to engage in clever analytic tricks for solving equations so that qualitative methods are obsolete. However numerical methods require vast amounts of data. They are expensive, and make generous funding necessary. They are good for extrapolation but less so for understanding. What is needed is a good combination of qualitative and quantitative mathematical methods.

5. With respect to present and future inquiry, how can the most important problems concerning evolutionary theory (or evolutionary aspects of your field) be identified and explored?

a. Look for connections between phenomena that are usually kept separate by disciplinary, institutional, and theoretical barriers. The mapping between genotype and phenotype is being enriched by dynamic systems theory which shows that small changes in parameters can produce major changes in outcome. The demonstration of a role for symbiosis in evolution places in the agenda the conditions under which alien organisms can survive and be incorporated into hosts as part of the genome.

b. Consider questions of what hasn't evolved. Why are there no co-enzymes that require lead, cadmium or aluminum? Why have no organisms evolved wheels? Can we experimentally select population to transcend the adaptive niches of their genera? (Once, long ago, Leigh Van Valen and I proposed a long term project in macro-evolution, selecting Drosophila to be aquatic (outside the family Drosophilidae but possible for insects. It remained in the bin of bright ideas, but Bruce Wallace picked it up and managed to select for aquatic larvae.)

c. Map "niche space", the ease of transitions between one and another mode of life. Some groups probe the limits of their adaptive zone repeatedly and spill over into other ways of life while other groups seem to be caught in adaptive prisons and show only limited ranges. In agricultural practice, natural enemies are often introduced to control pests. But we do not select to adapt already present parasitoids to adapt to new hosts. In epidemiology, the age of antibiotics may turn out to be a rather brief successional stage in our relations with the microbial world. We are running out of new ones that are sufficiently novel to thwart the parasites. But what if we used an evolutionary strategy? Suppose we treated severe cases of an infection curatively and mild cases palliatively. Could we select pathogens for reduced virulence and eventually toward commensalism?

c.More abstractly, we could apply dialectical principals to ask questions:

i. The truth is the whole. A problem has to be posed big enough for an answer to fit.It is easier to start too big and justify reducing the problem than to begin too narrowly and be dragged kicking and screaming to see the bigger picture.

ii. There will always be surprises. We have to be able to ask,

"But what if we're wrong?" And in particular to question the certainties of science. Students should be able to speculate on how might the second law of thermodynamics be overthrown, not to overthrow it but to emphasize that all theories have half-lives.While the main thrust of research has to be along lines that are judged to be plausible, there is the problem that the judges are the people who created the field as it is and are less likely to appreciate major departures. Therefore I announce the as yet unfunded Levins grants consisting of a hammock and a few cases of beer, awarded to promising investigators on the hope that they will come up with something interesting.

iii. Things are more connected than we expect. Therefore it is a good exercise to ask students for possible connections between seemingly unrelated phenomena. How does the ability of wheat plants to take up nitrogen affect the economic independence of women?

iv. Things are the way they are because they got that way. They haven't always been so, are not everywhere so, need not be so forever. This divides into the two questions, why are things the way they are instead of a little bit different, and why are things the way they are instead of very different. The first is the question of homeostasis (self-regulation). The second is the question of evolution, development, and history.

v. A "thing" is a snapshot of a process, preserved by a temporary balance of opposing forces long enough to earn a name.

vi. Apply all these tools to ourselves and the state of our field. The "scientific method" works pretty well to avoid idiosynchratic error: wash your glassware; don't divide by zero; every experiment needs a control; the patients and the administrator of pills should not know who gets the placebo. But it does little to correct the shared biases of our field, the "truths" which seem self-evident are never questioned. In order to escape the determination by our own narrow communities, we have to step outside. "Outside" might mean a different discipline, the same discipline in a different culture, or non-academic communities such as farmers or activist environmental groups. We can never be free of having a point of view but we can become more aware of our biases.

13
Elisabeth A. Lloyd

Arnold and Maxine Tanis Chair of History and Philosophy of Science, Professor of Biology, Adjunct Professor of Philosophy, Adjunct Faculty at the Center for the Integrative Study of Animal Behavior, Faculty in the Cognitive Science Program, Affiliated Faculty Scholar at the Kinsey Institute for Research in Sex, Gender, and Reproduction
Indiana University, USA

---

1. Why were you initially drawn to discussions and research on evolution (or evolutionary aspects of your field)?

I've been interested in evolution since I was six or seven and we got the Time/Life book, *Evolution*, and I saw the pictures of evolutionary changes and transformations. It was my favorite book. I came from a scientific family, and was exposed to museums, books, and scientific conversations when I was growing up. My dad was a mathematician at Bell Labs, where my mom also worked as a mathematician until she raised our family. I loved biology, and when I was in high school, I told people I wanted to be a marine biologist. But when I went to Queen's University in Canada in 1974 as a freshman and signed up for Biology 101, which I was really excited about, I was told by my academic advisor that it would be too hard for me, so I was un-enrolled and transferred over to a geography major with no biology. I ended up dropping out at the end of that year. When I went back to college at Boulder, Colorado, I signed up for Intro to Biology with no incident.

I got into philosophy through the back door. I had a terrific time in an honors course at Boulder, "Human Nature," and was already enrolling in an independent study with the professor for the following term when I discovered that he was, in fact, a philosophy professor. This mentor, Gary Stahl, eventually guided me towards writing my own major, "Science and Political Theory," in the Experimental Studies Department, so I could combine my pre-med science focus with my interests in political theory, continental philosophy, and intellectual history. I discovered philosophy of science my last year at Boulder, through reading Feyerabend and Kuhn, and wrote my honors thesis on Kuhn and experimental psychology and neuroscience.

But I was in for a nasty shock. When I got to Princeton's History and Philosophy of Science Program in 1980, in which I was enrolled in the Philosophy Department for a Ph.D., I discovered that philosophy of science really meant philosophy of physics. I was explicitly told that there was "no such thing as philosophy of biology." I was deeply inspired in my first year, though, by a course I was taking in the History Department, co-taught by Gerald Geison and Martin Rudwick, on the Darwinian Revolution. It was through my research in this course in 1981 that I came up with a paper to give to the brand new professor in the Department, Bas van Fraassen, for a course in Philosophy of Science. In the paper, I struggled to articulate the idea that Darwin seemed to have model outlines in his theory of natural selection, and that much of his writing, and especially his evidence, could be made sense of if we related it to this idea of models. I argued that this approach made more sense than the axiomatic view expressed in Michael Ruse's fascinating 1979 book, *The Darwinian Revolution*. Van Fraassen then took me out to lunch to discuss the paper, and mentioned gently that I might want to take a look at his new book, *The Scientific Image*, and what it has to say about models. I bought the book, and was so relieved to find that, lo and behold, here was a theory of models that I could use to articulate my idea about models in Darwin. So I wrote up the paper in terms of van Fraassen's "semantic view" of theories, and it was published in 1983 in *Philosophy of Science*, after a revision to take account of the fact that, meanwhile, Kitcher had published his syntactic account of model outlines. My Darwin paper gave me an entré to discuss philosophy of biology with others in the field (which did, indeed, exist), and I am indebted especially to David Hull, William Wimsatt, Margerie Grene, Philip Kitcher, and Lindley Darden for their early encouragement.

A big break came when I heard that Princeton had just instituted an exchange program with Harvard and some other schools, and I applied (over the graduate advisor's objections) to study in the Genetics Department at Harvard in Richard Lewontin's laboratory. I was very privileged that he accepted me as his student. I took a graduate seminar in 1983, "Problems in Evolutionary Theory," from Lewontin, Stephen Jay Gould and Pers Alberch, where I learned more about group and species selection, topics I would write about for some years. Since I was by then working on a formal analysis of evolutionary theory, especially selection theory, it seemed like an important challenge to figure out how selection theory was translated between biological levels, such as in group selection or species selection. I read Wimsatt's work on units of selection, which seemed basically right to me (and which I later defended against Elliott Sober's criticism), and based my analysis in my thesis on a modification of Wimsatt's analysis. The struggle to articulate a criterion for the entity that was selected in the natural selection process eventually led me to see that there is more than one question at stake in the debates, when biologists speak of or search for a "unit of selection."

It was also in graduate school that I began research on what much later became my book on explanations of the evolution of female orgasm. I was winding down one evening with a sister philosopher from out of town. In the context of a wide-ranging discussion, she asked me, "What is the function of the female orgasm, anyway?" I told her I'd look it up, and get back to her. I thought it would be like any other adaptation, and I could look up the account, just like "why we have opposable thumbs." I discovered, however, that the accounts that I looked up conflicted with what I knew about sexology evidence regarding variation and frequency in female orgasm. I searched further, and found an account which did not conflict with the sexology evidence, but rather, made sense of it: the account was based on the idea that the female orgasm is a developmental correlate of a trait actively selected in men, much like the male nipple is for the trait actively selected in women. The other accounts struck me as weakly supported and male-centered, but I set them aside. It was only later, when Evelyn Fox Keller asked me if I knew of any case studies of androcentrism (male-centered-ness) or sexism in science, that I remembered this case. I was invited to a conference to discuss such cases, and did more research on it to prepare for the conference, thus finding even more problems with the evidence. It did eventually seem to me, though, that the chief problem with the accounts was that they assumed that the trait of orgasm was an adaptation without anything to back this up, and even with some evidence indicating otherwise.

2. What does your work reveal about biological evolution (or evolutionary aspects of your field) that other academics, citizens, philosophers or biologists typically fail to appreciate?

My first paper, which argued that Darwinian selection should be understood in terms of models, rather than the logical positivist's axiomatic view of theories, was presented against a background in which it was taken for granted that theories should be presented, formally, in a law-like and axiomatic structure, as Michael Ruse sketches in his book on the Darwinian revolution (1979). I followed up the paper on Darwin with a paper on population genetics in terms of the alternative, models-based semantic approach to theory structure. The wide acceptance of the axiomatic view was partly due to the hard work of Mary Williams and J.H. Woodger (Williams 1970), who had carefully presented a partial axiomatic analysis of evolutionary theory earlier. So when I gave the first presentation of my models approach to evolutionary biology at the International Congress of Logic, Methodology, and Philosophy of Science in 1983, it was met with considerable scepticism. At that time, Bill Wimsatt and Morton Beckner were associated with thinking in terms of models, as was John Beatty, the first to promote a specific application of the semantic view to evolution. Ron Giere had also used evolution as an example of the semantic view. Robert Brandon and Ken Schaffner were friendly to a models-based approach, as well as another van Fraassen student, Paul Thompson, followed by James Griesemer, a Wimsatt student. But not that many philosophers of biology were interested in discussing biology in terms of models, as can be seen by examining the papers and conference proceedings at the time. Fortunately, times have changed. Now, much of the program at the International Society of History, Philosophy, and Sociology of Biology is imbued with model-talk and thinking in terms of models.

One aspect of analyzing evolutionary theory in terms of models that received much less attention from other authors until quite recently (with the exception of the writings of Wimsatt, especially), was the issue of confirmation and testing. My first paper concerned how Darwin brought evidence to bear on selective accounts, and emphasized the importance of independent empirical support for aspects of his model, which can be the chief form of support for some evolutionary models. In other words, empirical evidence and success for evolutionary models often comes most importantly in the form of support for its assumptions, rather than mostly in confirmation of its predictions or model outcomes. This also applies to population genetics models and ecological models, as, for example, ecologist Edward Rykiel later emphasized in 1996. The section on confirmation of evolutionary and ecological models that was part of my dissertation was published separately later. It was also reincorporated when my expanded dissertation was published as a book, *The Structure and Confirmation of Evolutionary Theory*, 1988 (republished by Princeton University Press, 1994). The book was a finalist for the Lakatos Prize in 1990.

Another significant part of my analysis of confirmation of evolutionary models concerned the variety of evidence, that is, the presence of several simultaneous but different sources of evidence for a model. In my paper, I use the real case of a biogeographical species theory being tested in tiny mangrove islands off the coast of Florida. The theory holds that there is a relation between species density and distance from the mainland, as well as island size. I claimed that not only do we need to test the parameter space of the model, as a type of variety of evidence that adds confirmation to the model, but that variety of evidence offers a novel type of confirmation. Specifically, the confirmation value of a single case of fit of the theory would improve if there were a variety of evidence from an array of cases. (This is different from the claim that the entire theory fits the array of cases – this claim concerns only the single case.) In other words, the theory applied to one case would be better confirmed if it were shown to apply to other cases. This seems counterintuitive, since the goodness-of-fit statistics for that one case do not change at all, in the context of having other cases fall under the umbrella of the theory. But in 1991 Brian Skyrms ran a hyper-Carnapian model of this claim (i.e., modelled it using a Bayesian framework), and found that the probability of the original hypothesis was higher. Thus, it's almost always true that having a variety of evidence supporting a theory increases the probability of a single application of the theory.

Much of my earlier work focused on various aspects of the challenges faced by representing levels of selection. It was out of discussions arising in the Harvard seminar, "Problems in Evolutionary Theory," that I realized that the requirement for emergent properties in species selection was too strong. Quite simply, we don't require that selection at the organismic level produce adaptations in order to say that selection is operating; it could simply be weeding-out selection, for instance. So why should we require an equivalently strong requirement for species level selection, i.e., that it produce adaptations at the species level, in order to say that species level selection is going on? (The term "adaptation" in these discussions involves something like an engineering notion; some improvement in design or function is attributed.) I saw the requirement of adaptation at the species level as an error in the position that Gould had published with Elisabeth Vrba in their 1986 paper. I wrote up this analysis in my 1988 book. I had by that time convinced Gould of the critique, but our joint paper on the topic, proposing an example from palaeontology, wasn't published until 1993 (with a follow-up in 1999). Our general approach to analyzing species selection has been endorsed both by leading palaeontologists Gene Hunt and David Jablonski (2006) and more generally by theorists Jerry Coyne and H. Allen Orr, in their book on speciation (2004).

By the late 1980s, I had thus realized that the different parties in several of the units of selection debates were pursuing subtly different questions. My thinking applied and expanded the earlier insights of Hull and Brandon, namely, that the evolutionary process involved both interactors – entities that interact with their environments as wholes in the selection process such that replication is differential – and replicators – entities of which copies are made in a selection process. To the interactor and replicator questions being pursued in the units of selection context, I added the questions, 'what is the beneficiary?', and 'what is the unit possessing or manifesting the adaptations?'. The issue of the beneficiary of an evolutionary or selection processes comes up repeatedly in discussions of genic selection or the "gene's eye" point of view. Their approach tends to ask, "who benefits in the long term from this selection process?", and their answer tends to be: genes or genetic elements. The issue of the manifestor-of-adaptations, on the other hand, is distinct from the interactor issue, as I was at pains to point out in the debate on species selection, because we can have an entity being selected at a level but not evolving a distinct adaptation or engineering design.

The biggest place where the manifestor-of-adaptation and interactor question became combined, though, was in the group selection debate. There, it had become customary, due mostly to the work of John Maynard Smith and George Williams, to search for group selection via searching for adaptations at the group level. For example, group selection was thought to require that there be a group-level adaptation such as altruism, in which the group fitness or interest conflicted with organismic fitness; arguments against this possibility were thus believed to be arguments against the possibility of group selection overall. Hence, the implication was that the unit of selection for Maynard Smith (and Dawkins), in the context of the group selection debate, required both that the entity was an interactor and that it possessed an adaptation at the group level. Other researchers working on group selection coming out of the tradition of Sewall Wright, such as Michael Wade or Charles Goodnight, required only that the group behave as an interactor; no adaptation at the group level was required. Thus, when Maynard Smith disagreed with Wade about group selection, they were clearly talking at cross-purposes because they actually defined a unit of selection differently. This same kind of cross-purpose talk also occurred in the species selection debate, as mentioned above. One side required both an interactor and the possession of an engineering adaptation at a level, while the other side required only an interactor at the species level. And so in my 2001 article on units of selection for the Lewontin Festschrift, this simple set of distinctions could be put to work to sort out some disagreements. Maynard Smith gave the article a thumbs-up in *Evolution* when it was first published (an expansion is now available online at the *Stanford Encyclopedia of Philosophy*), and it's been reprinted elsewhere.

My basic analysis of the units of selection problems first appeared in *Keywords in Evolutionary Biology* (1992), a collection of essays I edited with Evelyn Fox Keller. Evelyn conceived the book, and we solicited essays on the multiple meanings of key concepts in evolutionary biology from leading researchers in biology as well as history and philosophy of biology. I felt very privileged to gain the cooperation of such a stellar group of authors, and to be co-editing the book with Evelyn. We hoped the book might serve as both a resource for biologists and people in the HPS field, as well as a spur to future research in the area. (A Keywords volume in Developmental Biology was published several years ago, edited by Brian Hall and Wendy Olson.)

At the same time, the work of Leda Cosmides and John Tooby was brought to my attention by geneticist Marcus Feldman (Stanford). I developed a critical analysis of their claim that there was an evolved brain module for detecting cheating on social contracts that explained a puzzle in psychological testing. The puzzle was that people don't perform well with an abstract "if-then" statement, such as "If A then B" (as anyone who has taught logic will tell you), but their performance improves dramatically if certain social contexts are given, for example, if the case involves checking IDs for drinking alcohol (If you drink alcohol, then you must be 21). Cosmides and Tooby claimed to show that this phenomenon is explained by an evolved module in the brain for detection of cheating on a social contract, in opposition to a learning model. In my analysis of the paper, I argued that they had supported no evolutionary claim at all, and had failed to eliminate the competing learning model, as they had claimed. That paper was eventually published, followed by a joint paper with Feldman regarding the dangers of evolutionary psychology's approach to evolutionary genetics. Evolutionary psychologists, we found, appealed to only a small corner of simplified genetics theory, while ignoring portions of the theory that would have been more appropriate to the task. As a result of relying on inclusive fitness theory for their genetics, evolutionary psychologists also had difficulties explaining variation, and they made assumptions about optimization and adaptation that would also be inappropriate and undefended in many cases. With my subsequent critique of Randy Thornhill and Craig Palmer's book on the evolution of rape, I added to a growing chorus of critics of evolutionary psychology.

My book, *The Case of the Female Orgasm: Bias in the Science of Evolution* (2005) examines all 21 theories that I could find in the literature offering some account of how human female orgasm evolved. The largest debate in the literature — although it was more assumed than debated — was whether or not female orgasm was an evolutionary adaptation. I approached the problem by laying out the standard evolutionary requirements for showing that a trait is an adaptation, and then considering the accounts given for female orgasm. I noted the conflict between 10 of the accounts and the sexology literature, a conflict that I chalked up to an androcentric (male-centered) bias. This was because they were assuming that women responded to intercourse the same way that men do – with an orgasm pretty much every time. This is not a correct assumption about women, however. Only 25% of women have reliable orgasm with intercourse, 30% rarely or never have orgasm with intercourse, and a further 10% of women never have orgasm at all by any means (2005, p. 36). This discrepancy struck me as a pretty significant empirical problem for these accounts.

There was another problem with these explanations, however, one that they shared with 9 of the other accounts of female orgasm, namely the assumption that female orgasm was itself an adaptation. The problem is that women with more orgasms don't seem to have more (or better) babies; that is, no fitness correlate with orgasm has been found, and thus there is no evidence that orgasm is an adaptation. It may once have been, but there's no evidence for that, either, given the very wide spread in variation in the population that we have in orgasmic frequency, which is likely an indicator of a lack of selection.

As I discovered when exploring the various explanations for female orgasm, the evolutionary account that is best supported at the present time is the developmental byproduct account, under which female orgasm is understood as a developmental correlate of strong selection on male orgasm. Thus, the case raises big questions: why was the best explanation on the evidence repeatedly rejected in favor of theories that were clearly inferior? Why did so many of the explanations conflict with the sexology evidence, and why didn't that bother more of those involved in the debate? How did a clearly problematic explanation, with failing statistical support — the "upsuck" theory, in which sperm were moved through the reproductive tract by a sucking motion of the uterus upon orgasm — manage to become the "fact" of the matter about female orgasm? It had been contended for some years by feminist philosophers of science that sometimes social bias plays a role in the practice of science – not just in the formulation of problems, or in the funding of scientific projects, but also in theory evaluation and consideration of evidence. The evolution of female orgasm is a clear example of such a case. On a different note, there had been a struggle in evolutionary biology between those who pursue adapationist explanations for traits more or less exclusively and those who entertain a wider range of explanations. This case is one in which an adaptationist approach appears to have been the motivation for accepting a clearly substandard explanation over its nonadaptive rival. Developing understanding and insight about the ways that bias — either androcentric or adaptationist — can operate in science is one of the chief aims of the book.

3. What, if any, practical and/or social-political and/or moral obligations follow from your work on evolution?

I have been concerned with the natural variability that occurs in the human population, how that's understood theoretically, and its potential social consequences. (One might think that this interest in variability is not too surprising, given that I studied with Lewontin, and Lewontin studied with Dobzhansky, who emphasized the value of variability in evolution.) In an early paper on normality and variation in relation to the Human Genome Project, I argued that genetics itself does not and will not give us a workable notion of health or normality. Even with a genetic decoding of an individual, we would ultimately define a variant gene as "abnormal" or "diseased" only if it interfered with some notion of proper functioning. The notion of proper functioning, in turn, arises from a medical model at the organismic level, and the standards of good health are socially negotiated. (For example, think of height or degree of body fat as markers of good health over time and across cultures.) In addition, both developmental and population genetical models are necessary to fully understand and describe the processes involved in genetic disease, in addition to the genetics and medical models. Each of these models has a separate and distinct meaning of "normal" and "abnormal" attached to it, but often these distinct notions of normality get interchanged during discussion of disease states or genetic differences. This lack of rigorous thinking about the models is especially risky for us when the traits under discussion are of special social importance.

The issue of normality arises also in my book, *The Case of the Female Orgasm*. Evolutionary accounts of human traits tend to get interpreted as cultural norms, especially if they concern socially relevant traits. Thus, if female orgasm naturally evolved to help us maintain our pair bonds, and the evolutionary account describes women "naturally" having orgasm with intercourse nearly every time intercourse occurs, then women who deviate from that behavior will be seen as "abnormal", "unnatural" or "deficient", as indeed was the case when these pair bond accounts were written. The key issue that arises with these pair bond accounts, is not simply that they conflict with over 80 years of evidence from sexology (i.e., the facts cited above about the relatively low rates of female orgasm with intercourse) – they also serve as the foundation of social norms. That is, they naturalize particular expectations about how women are assumed to respond sexually, by providing a scientific foundation for that response. Since, counterintuitively, there is no evidence that more orgasms lead to more babies, the competing, byproduct theory has the strength of being non-adaptive, and reflects just this non-association. In other words, it unlinks reproduction from female orgasm, and thus potentially from female sexuality.

Many feminists have been striving for many decades to unlink their reproductive function from their identities as women, including their sexual identities, so this evolutionary theory would seem to be a welcome gift. Making female sexuality autonomous, from both male sexuality and from reproduction, as the byproduct theory appears to do, leaves the field wide open for social discussion regarding sexuality. It thus provides a very different sort of groundwork or foundation for narratives about female sexuality than, for instance, the old pair bond theories did.

The other key feature of the byproduct (or fantastic bonus) account is that it makes sense of the wide variation that we see in women's response to intercourse. (I was moved to start calling it the "fantastic bonus" account because of feminist objections that the "byproduct" account made orgasm "sound like a can of spam"...) If the trait is not under selection, a wide variation in response might be expected, just as we find. (It is theoretically possible for sexual selection theory to produce the variation we see, but not within realistic values for the selective forces.) Thus, the woman who never has orgasm at all is just as "normal" as the woman who has orgasm all the time with intercourse – there is really no distinction between them under the byproduct view, since none are any more "functional" than the other. I learned from personal accounts of women who have read my book that being included in the range of "normal" women is extremely meaningful to them.

Thus, although we may not, ourselves, draw social and value judgements from the science, others inevitably will, and we must be prepared to think through the consequences. The statistics that I collected and summarized in my book about the relatively low rate of orgasm with intercourse, and the wide variation among women in their responses, have been discussed or quoted in women's magazines and health magazines. The broad dissemination of these features of the research indicates to me that there is, indeed, intense interest in the variation in female response and its explanation in biology. The old explanations for the lack of orgasm in intercourse, such as blaming the male for ineptitude or the female for uptightness, are, well, old. Facing a scientific summary stating that some women are, basically, just right the way they are, was perceived as different or refreshing. The autonomous sexuality of women thus gets showcased through the best-supported evolutionary approach available today.

4. What do you see as the most interesting criticism against your position in the biological or philosophical discussion of evolution?

Some of the most philosophically interesting criticism of my work has arisen from my recent work about genic selection and pluralism, especially a 2005 paper, "Why the gene will not return." C. Kenneth Waters (2005) wrote a rebuttal to the piece, and others are involved in the general discussion in interesting ways, such as Robert Wilson, Elliott Sober, Kim Sterelny, Paul Griffiths, and Peter Godfrey Smith. The sort of pluralism that I'm concerned about is articulated in Sterelny and Kitcher's very popular 1988 paper, "The return of the gene," and in the paper they co-authored with Ken Waters in 1990. There, the claim is that all selection processes can be represented with the gene as the focal unit of selection, i.e., with the gene as the central causal interactor. It is this claim that I was interested in rebutting in my 2005 paper. By showing that the genic models were completely formally derived from the hierarchical models, I argued that there is no room for these models to have acquired the primary genic cause that is needed by the pluralist account. A derived model cannot have more causes than the original model had. (Note that this still leaves room for genuine cases of genic selection, which would show up in the original hierarchical model, and would also appear in the genic derived model.)

I had previously written about and accepted multiple types of models representing the same system in nature (in some cases) in my first book. It is not pluralism per se that I have a problem with; in fact, I am a pluralist, under a moderate definition.

Waters' rebuttal provides good insight into some of the issues with pluralism. He describes "Plurality of Empirical Reasonable Models" (PERM) situations, in which more than one empirically adequate model can account for a selection process, given what we know. As Waters approaches it, PERM situations are so complicated that there are multiple ways to disentangle them, and he remarks that it is not clear whether multilevel theories can be developed to sort out causes at distinct levels. On his view, there is a plurality of ways to sort out causes at multiple levels. Waters emphasized, following Reichenbach, that it is the philosophers' job to investigate the consequences of scientists' choosing to follow either pluralism or monism regarding genic or multiselectionist models. This investigation includes exploring different ways to identify causes in evolutionary models. Waters uses Jim Woodward's theory of causation and Helen Longino's notion of parsing causes to sort things out and defend his view of pluralism. Rob Wilson also defends pluralism, although with a different set of definitions of pluralism and causes, etc., and with an emphasis on defending hierarchical evolutionary models. Peter Godfrey Smith defends pluralism on other grounds. The discussion continues to involve foundational questions such as the nature of cause, the interpretation of models, and the nature of the entities and processes. Waters notes that it is important to clarify two issues that are often opposed to pluralism by its opponents: realism and conventionalism. For instance, he is a self-described "tempered realist" about selection models, believing that some of them do offer correct or true descriptions of selection processes, yet he is also a pluralist.

One of the points of tension I have with the pluralists involves the role of history of science in doing the philosophy of science, a perennial problem for philosophers of science. What credit should be given to the hierarchical theory for discovering information about higher-level causes? The pluralists say: none of that is relevant. On a related point, does the use of methodology from hierarchical theory mean that one is committed to interpreting one's results using hierarchical theory? (Clearly not, but what if there is no methodology from the lower-level theory that can do the job at all?) This kind of question brings up both new and old problems concerning theory reduction, involving both epistemology and metaphysics. These sorts of problems, combined with other issues about the nature of selection processes and the entities playing key roles in those processes, as well as how to interpret models, continue to challenge philosophers of biology, even after some of the key debates involving units of selection have been resolved.

5. With respect to present and future inquiry, how can the most important problems concerning evolutionary theory (or evolutionary aspects of your field) be identified and explored?

The best way to identify and explore the most important problems concerning evolutionary theory is to talk with and hang around evolutionary biologists. This involves going to their talks and seminars, starting reading groups and enticing them to attend, working with their students, inviting them to serve on your (or your students') masters, qualifying, and dissertation committees, and so on. It may also involve, if they permit this, attending their laboratory meetings. Attending the big Evolution Society meetings is also useful.

Some of the emerging or exciting problems in and around evolution today include issues surrounding speciation and species identity – there is much theoretical and experimental work being done right now in how species come to arise out of other species, and arguments about how to tell when this is happening. It is fertile ground for philosophical work. This is related to the broader question of evolvability. Why does one lineage evolve and ramify and another does not? What about a species enables it to respond to environmental stressors and to evolve adaptations, while another species becomes extinct? These and other questions surrounding evolvability involve several biological fields and are hot topics in biology right now, needing some conceptual clarification.

Another emerging area needing philosophical attention is the biomedical trends towards individual genomics and diagnostics. The historical antecedents and ramifications of these practices are fruitful spots for philosophical scrutiny, especially given their social significance. On a more theoretical note, philosophers could explore the impact of indirect genetic effects on various areas in evolutionary biology, including kin selection, social evolution, niche construction, or community genetics. In all of these cases, indirect genetic effects influence the usual, informal characterization of the "nature/nurture" paradigm, and deserve elucidation and analysis.

References

Coyne, J. A. and H. Allen Orr (2004) *Speciation*. Sinauer: Sunderland, MA.

Jablonski, D. and G. Hunt (2006) "Larval ecology, geographic range, and species survivorship in Cretaceous mulloscs: Organismic versus Species-level Explanations," *American Naturalist* 168: 556-564.

Lloyd, E.A. (2001) "Units and Levels of Selection: An anatomy of the Units of selection debates" in *Thinking about Evolution: Historical, Philosophical, and Political Perspectives*, edited by R.S. Singh, C.B. Krimbas, D.B. Paul and J. Beatty. Cambridge University Press: New York, pp. 267-291.

"Normality and Variation: The Human Genome Project and the Ideal Human Type," in *Are Genes Us? The Social Consequences of the New Genetics*, ed. Carl F. Cranor, Rutgers University Press, 1994, pp. 99-112.

Lloyd, E.A. and M.W. Feldman (2002) "Evolutionary Psychology: A view from evolutionary biology," *Psychology Inquiry* 13 (2): 150-156.

Lloyd, E.A. and S.J. Gould (1993) "Species Selection on Variability," *PNAS* 90: 595-599.

Wallen, K. and E.A. Lloyd (2008) "Inappropriate comparisons and the weakness of cryptic choice: a reply to Vincent J Lynch and D. J. Hosken," *Evolution and Development* 10 (4): 398-399.

Waters, C.K. (2005) "Why Genic and Multilevel Selection Theories are Here to Stay," *Philosophy of Science* 72 (2): 311-333.

Williams, M.B. (1970) "Deducing the consequences of evolution: A mathematical model," *Journal of Theoretical Biology* 29: 343-385.

14
Stuart A. Newman

Professor of Cell Biology and Anatomy
New York Medical College, USA

---

1. Why were you initially drawn to discussions and research on evolution (or evolutionary aspects of your field)?

I was trained in the physical sciences and was encouraged by my doctoral supervisor at the University of Chicago, chemist Stuart A. Rice, to take a systems approach to the theoretical analysis of complex chemical systems. The late 1960s were early days in the analysis of complexity, and many now popular concepts – multivariate oscillations, deterministic chaos, reaction-diffusion patterns – were recently developed or just coming into existence. New mathematics was needed to conceptualize these problems and the 1970s saw a burst of activity in this field. With the rapid rise of high-speed computing that also occurred during this period it began to be possible to actually test the validity of models for the behavior of complex systems and predict the outcomes of such behavior.

It turned out that in chemical systems with many components the outcomes (what biologists call "phenotypes") tended to fall into a small number of fairly simple categories. These came to be known as "dynamical attractors" of the systems. More specifically, for chemical systems they are modes of behaviour in which concentrations of all the components are either unchanging but maintained away from chemical equilibrium by fluxes of matter and expenditures of energy ("steady states"), oscillatory in time ("limit cycles"), or randomly traversing a confined region of the system's state space ("strange attractors"). This was a real lesson concerning the nature of matter: complex systems tended to spontaneously simplify, giving rise to levels of organization and phenomena unanticipated by examination of the systems' fundamental units. But these emergent modes of behavior and organization were only evident when the systems were viewed at appropriate scales – they were not generated by the units acting like clockwork machines or programmed computers.

It seemed to me that many-component chemical systems viewed in this fashion were models for living systems and their mysteriously coexisting stability and plasticity. Cell metabolism was an instance of this kind of system; as an undergraduate I was awed by the elaborate charts of intracellular networks of molecular transformations that decorated the biology professors' walls, replaced on a frequent basis by ever more complex versions. My Ph.D. thesis, "Biological Aspects of Complex Chemical Systems," was an attempt to model some general features of intermediary metabolism by the dynamics of enzyme networks. My approach was influenced by the early work on genetic switching networks by Stuart Kauffman,[1] who had recently joined the faculty at the University of Chicago and had agreed to serve on my doctoral committee.

As a graduate student, and later a postdoctoral fellow in Chicago's department of theoretical biology, I was extraordinarily fortunate to be surrounded by scientists who were pioneers in many of the major areas of theoretical biology.[2] The chemistry department during my time there had yearly visits from Ilya Prigogine, who with his colleague Gregoire Nicolis, then a research fellow in the Rice group, was breaking new ground in the physical theory of self-organization.[3] Theoretical biology had as sabbatical visitors Brian Goodwin,[4] a founding figure of what later came to be known as multiscale modelling of biological systems, and John Maynard

---

[1] Kauffman, S. A. 1969. Metabolic stability and epigenesis in randomly constructed genetic nets. *J Theor Biol*, 22:437-67.

[2] Among these were Stuart Kauffman, as mentioned, Jack Cowan (Cowan, J. D., and D. H. Sharp. 1988. Neural nets. *Q Rev Biophys*, 21:365-427), Leon Glass (Kaplan, D., and L. Glass. 1995. *Understanding nonlinear dynamics*. Springer-Verlag, New York), Richard Levins (Levins, R. 1968. *Evolution in changing environments; some theoretical explorations*. Princeton University Press, Princeton, N.J.), Richard Lewontin (Lewontin, R. C. 1974. *The genetic basis of evolutionary change*. Columbia University Press, New York), Vidyanand Nanjundiah (Nanjundiah, V. 1973. Chemotaxis, signal relaying and aggregation morphology. *J Theor Biol*, 42:63-105) and Arthur Winfree (Winfree, A. T. 2001. *The geometry of biological time*. 2nd ed. Springer, New York.)..

[3] Prigogine, I., and G. Nicolis. 1971. Biological order, structure and instabilities. *Q Rev Biophys*, 4:107-48.

[4] Goodwin, B. C. 1963. *Temporal organization in cells; a dynamic theory of cellular control processes*. Academic Press, London, New York.

Smith, the most esteemed evolutionary theorist of the period, whose unique array of interests also included developmental pattern formation.[5]

Although I had a surreptitious interest in evolution at this point, it was not on my own research agenda. Everything in biology that I had personally studied, and that drew me to the work of the scientists mentioned above and others working in the same vein, told me that complex behaviours and structures arose by self-organization. Certain properties emerged at particular scales due to inherent system dynamics, and the choice of which of a system's latent characteristics were realized was as often as not influenced by external parameters. Applied to organic evolution these views smacked of saltationism, orthogenesis and Lamarckism. Under the circumstances, which to be precise, were simultaneously the pinnacle of the Modern Evolutionary Synthesis[6] and the threshold of the era of molecular genetics of multicellular organisms,[7] my evolutionary ideas were a potential embarrassment, and I largely kept them to myself.

To elaborate on these heresies, saltationism is the idea that organismal phenotype can change suddenly from one generation to the next in a fashion that is very large, in comparison with the organism's usual range of phenotypic variation. Although the Modern Synthesis can tolerate the occasional saltation or jump, population genetics under neo-Darwinian assumptions held them to be exceptionally rare occurrences that would not have contributed much to the origin of species or higher taxa. By the 1970s, saltationism had attained a proscribed status in the form of the "hopeful monster," a term invented by the geneticist Richard Goldschmidt to refer to novel phenotypes that might arise in the space of a generation by the mutation of a single broadly acting gene (a "macromutation").[8] Apart from its deviation from the predictions of population genetics, the attribution of an evolutionary role for saltations violated a deep disposition of the Synthesis en-

---

[5] Maynard Smith, J. 1968. *Mathematical ideas in biology.* Cambridge University Press, Cambridge.

[6] Dawkins, R. 1976. *The selfish gene.* Oxford University Press, New York; Smocovitis, V. Betty. *Unifying Biology: The Evolutionary Synthesis and Evolutionary Biology,* Princeton University Press, 1996

[7] Davidson, E. H. 1976. *Gene activity in early development.* 2d ed. Academic Press, New York.

[8] Goldschmidt, R. B. 1940. *The material basis of evolution.* Yale University Press, New Haven.

capsulated in a famous statement by Darwin in the first edition of *On the Origin of Species,* "[i]f it could be demonstrated that any complex organ existed, which could not possibly have been formed by numerous, successive, slight modifications, my theory would absolutely break down".[9]

Orthogenesis is the doctrine that organisms change in preferred directions over the course of evolution. Although this idea is consonant with the fact that all material systems have inherent patterns of organization that may manifest themselves over time (as in the evolution of the chemical elements), and despite it having been approached in terms of the physical properties of living tissues by the biologists William Bateson[10] and D'Arcy W. Thompson,[11] orthogenesis was roundly rejected by such leading architects of the Synthesis as George Gaylord Simpson[12] and Ernst Mayr[13] as incompatible with this framework.

Lamarckism was the most scorned of all the heterodoxies, despite the inheritance of acquired characteristics having been embraced with increasing avidity by Charles Darwin himself in successive volumes of *The Origin*. First caricatured by the anti-evolutionist Georges Cuvier based on a passing comment about the hypothetical neck-stretching giraffe,[14] Jean-Baptiste Lamarck's considerably subtler physicalist and organism-environment interactionist views had been written out of the Synthesis in part because of the low degree of phenotypic plasticity of the fruitfly *Drosophila melanogaster*, whose biology was paradigmatic during the rise of the theory, but later also because of the taint of its ill-fated espousal by the government of the Soviet Union in the form of Lysenkoism. With the Cold War still in effect, the early 1970s were not a favorable period in which to counter the accepted evo-

---

[9]Darwin, C. (1859). "On the origin of species by means of natural selection, or, The preservation of favoured races in the struggle for life." J. Murray, London.

[10]Bateson, W. 1909. Heredity and variation in modern lights. In A. C. Seward (ed.), *Darwin and modern science*. Cambridge University Press, Cambridge.

[11]Thompson, D'A. W. 1942. *On Growth and Form*. 2nd ed. Cambridge University Press, Cambridge.

[12]Simpson, G. G. 1944. *Tempo and mode in evolution*. Columbia Univ. Press, New York.

[13]Mayr, E. 1974. Teleological and teleonomic: a new analysis. *Boston Studies in the Philosophy of Science*, 14:91-117.

[14]Burkhardt, R. W. 1995. *The spirit of system: Lamarck and evolutionary biology*. Harvard University Press, Cambridge, Mass.

lutionary narrative with doubts based on a still inchoate mixture of scientific taboos, some of which seemed suspiciously Communistic.

Genes, though vigorously rejected by some of my scientific mentors as exclusive causative factors of the biological phenotype,[15] continued to hold prime place as determinants of phenotypic change in standard evolutionary theory. The Modern Synthesis had as its default explanatory mode gradual population-level change by selection of genetic variants associated with incremental improvements in the phenotype, and none of the discoveries made during the great period of activity in genetics and genomics research in the 1970s and 80s disturbed that mainstream view. It is significant to note that Stephen Jay Gould, whose paleontological concept of "punctuated equilibrium"[16] was initially interpreted (including possibly by Gould himself) to support saltationist mechanisms of macroevolution, had entirely disavowed this interpretation by the end of his life. While his final treatise advocates a pluralistic view of levels of selection, it is forthright in asserting that all phylogenetic change, including the gaps in the fossil record recorded as punctuations, proceeds by incremental microevolutionary processes.[17]

My notion of cellular and developmental systems as plastic, though also constrained because of self-organization, was fortified by discussions with Stuart Kauffman at Chicago and Brian Goodwin, whom I joined as a postdoctoral fellow at the University of Sussex in the early 1970s. Richard Lewontin, whom I had known in Chicago, was on sabbatical at Sussex with John Maynard Smith during the same period, and I absorbed evolutionary concepts by listening in on their lively and sometimes heated discussions over departmental afternoon teas. The various influences on me did not always add up, and I began to increasingly voice my qualms about some features of the Darwinian synthesis. Toward the end of my time in Sussex John Maynard Smith commented, "I used to think you didn't get what we were saying. Now I think

---

[15] Lewontin, R. C., S. P. R. Rose, and L. J. Kamin. 1984. *Not in our genes: biology, ideology, and human nature.* 1st ed. Pantheon Books, New York; Levins, R., and R. C. Lewontin. 1985. *The dialectical biologist.* Harvard University Press, Cambridge, Mass.

[16] Gould, S. J., and N. Eldredge. 1977. Punctuated equilibria: The tempo and mode of evolution reconsidered. *Paleobiology*, 3:115-151.

[17] Gould, S. J. 2002. *The structure of evolutionary theory.* Belknap Press of Harvard University Press, Cambridge, Mass.

you just don't agree."

My transition from amateur to professional status as a student of evolution first required experimental engagement with an actual developmental system. Back in the United States, I received intensive training in the newly emerging understanding of gene organization and expression in multicellular organisms under the guidance of Eric Davidson and Fotis Kafatos in the Embryology Course of the Wood Hole Marine Biological Laboratory. Then as a research fellow at the University of Pennsylvania and later at the State University of New York at Albany, where John W. Saunders, Jr., the great limb embryologist was also on the faculty, I began a series of studies on skeletal pattern formation that has continued to the present. By observing the self-organizational properties of embryonic limb cells in culture I came to appreciate the role of cell adhesion and extracellular matrix (ECM) molecules, and diffusible activators and inhibitors of cell state switching, in pattern formation and morphogenesis. This provided the basis for a theoretical model developed in collaboration with the physicist H. L. Frisch in which the limb skeleton is shown to be potentially generated by the patterned induction of ECM by diffusible molecules (morphogens) acting as a local autoactivation-lateral inhibition regulatory network in a growing tissue domain.[18]

If the limb could arise from cells interacting with molecules which not had evolved specifically to produce limbs, by a balance of physical interactions in a context-dependent fashion, it could well have emerged evolutionary in a relatively abrupt fashion, without substantial genetic change in a non-limb-bearing predecessor. Incremental adaptive scenarios were not necessary if physics was added to the causal mix.[19] Around the time we were formulating our model several influential works appeared that began to dislodge the conventional wisdom concerning the role of adaptation in morphological change. Stephen Jay Gould was involved in two of these: (i) his book *Ontogeny and Phylogeny*[20] reviewed the concerted, large-scale changes that can be effected

---

[18] Newman, S. A., and H. L. Frisch. 1979. Dynamics of skeletal pattern formation in developing chick limb. *Science*, 205:662-668.

[19] Earlier on, William Bateson (see note 7) and D'Arcy W. Thompson (see note 8) had recognized this, and before them, Lamarck (Lamarck, J. B. 1809 (trans. 1984). *Zoological philosophy: an exposition with regard to the natural history of animals*. University of Chicago Press, Chicago.

[20] Gould, S. J. 1977. *Ontogeny and phylogeny*. Harvard University Press, Cambridge, MA.

in adult forms by small alterations at early developmental stages, and (ii) a seminal paper of his with Richard Lewontin[21] demonstrated the inevitability of some biological characters arising as non-adaptive side-effects of other evolutionary processes.

These contributions, for all their iconoclasm, remained tied to incremental change of the phenotype via the classic population-based, variational, genotype-phenotype connection. A little later, however, some contributions to a conference volume edited by John Tyler Bonner[22] and the appearance of new experimental and conceptual work by a young Catalan, Pere Alberch[23] expanded the discourse in evolutionary biology so as to incorporate developmental plasticity. This provided the foundation for the new field of evolutionary-developmental biology, or at least the non-gene-centric strain in which I began to participate.

My initial foray into evolutionary theory came in the non-traditional form of a conceptual review in the journal *Development*,[24] which I co-wrote with the physical biochemist Wayne Comper. Here we attempted to draw general lessons from developmental studies such as my own on the vertebrate limb in which physical mechanisms were purported to play an important role. It became clear from close analysis of the relevant cases that morphological motifs that were "obviously" generated by physical processes (multilayered frog gastrulae, segmented fruit fly embryos, and so forth) were actually produced in much more complicated fashions, by mechanisms that were undoubtedly built up over long periods of time. We proposed, however, that the complexity of the developmental mechanisms carried no implications concerning the *origins* of the resulting shapes and forms were similarly complex. From the disparity between the robust developmental processes of present-day embryos and the conditionality and plasticity of the "generic" physical mechanisms capable of producing the same outcomes, the idea emerged that morphological evolu-

---

[21] Gould, S. J., and R. C. Lewontin. 1979. The spandrels of San Marco and the panglossian paradigm. *Proc. Roy. Soc. London B*, 205:581-598.

[22] Bonner, J. T., Ed. 1982. Evolution and development. Springer Verlag, Berlin ; New York.

[23] Alberch, P., and E. A. Gale. 1983. Size dependence during the development of the amphibian foot. Colchicine-induced digital loss and reduction. *J Embryol Exp Morphol*, 76:177-97; Alberch, P. 1989. The logic of monsters: evidence for internal constraint in development and evolution. *Geobios*, 19:21-57.

[24] Newman, S. A., and W. D. Comper. 1990. 'Generic' physical mechanisms of morphogenesis and pattern formation. *Development*, 110:1-18.

tion was not typically, as Darwin had believed, a gradual process molded by uniform causes. Physical processes could act early and quickly, providing templates for the gradual accretion of genetic reinforcers, with the latter often obscuring the originating agency of the former.

2. What does your work reveal about biological evolution that other academics, citizens, philosophers or biologists typically fail to appreciate?

According to Ernst Mayr, "Nothing strengthened the theory of natural selection as much as the refutation, one by one, of all the competing theories, such as saltationism, orthogenesis, inheritance of acquired characters, and so forth."[25] It is easy to see why the first two of these are completely antithetical to Darwin's concept that complex organismal structures arise by increments, based on adaptation to external conditions, with no preferred directions of change. The third competing theory does not specifically violate the precepts of Darwin who, as mentioned above, was increasingly open to a role for inheritance of characteristics acquired during an individual's lifetime. It conflicts rather with neo-Darwinism: the melding of natural selection with population genetics that is the core of the Modern Synthesis. For animal species, at least, the barrier between body cells and the germline postulated by August Weismann[26] shortly after Darwin's death is an unquestioned component of the Synthesis, one that is considered to preclude Lamarckian mechanisms.

The evolutionary framework my colleagues[27] and I have developed over the past two decades incorporates all three of the prescribed concepts, though perhaps not exactly in the way that Mayr and the other supporters of the Synthesis have decried. In fact, our model does not dispute the applicability of the neo-Darwinian model to classical cases of microevolution such as the reshaping of finch's beaks[28] and other cases of the change in size and shape of

---

[25] Mayr, E. 1982. The growth of biological thought: diversity, evolution, and inheritance. Belknap Press, Cambridge, Mass, p. 840.

[26] Weismann, A. 1892. Das Keimplasma; eine Theorie der Vererbung. Fischer, Jena.

[27] Primarily, the evolutionary-developmental biologist Gerd B. Müller, and more recently a graduate student, Ramray Bhat.

[28] Müller, G. B., and S. A. Newman. 2005. The innovation triad: an EvoDevo agenda. *J Exp Zoolog B Mol Dev Evol*, 304:487-503.

defined structures.[29] While we strongly question the relevance of this mechanism to the origination of body plans and morphological novelties,[30] our alternative explanation for macroevolutionary change, unlike those of creationists and advocates of "Intelligent Design,"[31] is entirely naturalistic.[32]

Our view starts with the observation that present-day organisms result from developmental processes that are highly "canalized." This term, introduced by C. H. Waddington,[33] refers to the evolved property of staying on track despite the inevitable presence of internal biochemical noise, external perturbations and genetic polymorphism. Under these circumstances there will be a close relationship between genotype and phenotype, and in general (because of multiple fail-safe mechanisms) the phenotype will only change marginally when the genotype is altered. These are the conditions required for gene-associated variations with small effect subject to natural selection.

The major animal body plans, however, have been in existence for a least half a billion years. This implies that during the intervening period much of the evolution of the relevant developmental pathways, rather than generating new body plans, consolidated the realization of the existing ones. Therefore, the earlier the evolutionary stage the more easily would developmental pathways be thrown off track by either mutation or environmental change. Another way of stating this is that the mapping between genotype and phenotype was less rigid in the less canalized forms of the past than at present. Gene alterations and environmental changes would typically have had large effects on the phenotype. But these large effects (i.e., saltational changes) would not have been chaotic or haphazard. Because organismal characteristics were (by our hypothesis), originally moulded by self-organizing processes (rather

---

[29] Klingenberg, C. P., L. J. Leamy, and J. M. Cheverud. 2004. Integration and modularity of quantitative trait locus effects on geometric shape in the mouse mandible. *Genetics*, 166:1909-21.

[30] Darwin (1959; see footnote 8); Grant, P. R., and B. R. Grant. 2005. Darwin's finches. *Curr Biol*, 15:R614-5.

[31] Behe, M. J. 2007. The edge of evolution: the search for the limits of Darwinism. Free Press, New York.

[32] Newman, S. A. 1994. Generic physical mechanisms of tissue morphogenesis: A common basis for development and evolution. *Journal of Evolutionary Biology*, 7:467-488; Newman, S. A., and G. B. Müller. 2000. Epigenetic mechanisms of character origination. *J. Exp. Zool. B (Mol. Dev. Evol.)*, 288:304-17.

[33] Waddington, C. H. 1942. Canalization of development and the inheritance of acquired characters. *Nature*, 150:563-565.

than by the usually supposed genetic programs), a common set of alternative phenotypes would emerge as a result of environmental or mutational change, namely, the limited, inherent attractors of the developmental system. (The phenomenon of the "phenocopy" that led Goldschmidt to his hopeful monster theory and Waddington to his concept of canalization is illustrative of this principle.) The tendency to assume preferred morphologies is, by definition, orthogenesis.

The mechanism of phenotypic change provided by the Synthesis requires a series of gradual steps, and for this the original organisms must be relatively well adapted to their environment. Organisms that are morphologically very different from members of their originating population would not qualify for evolution via this incrementalist mechanism. But if they have already been changed significantly by some other process they would not need to conform to the Darwinian paradigm to account for the change. The only question would be how they would fit into the ecosystem, since their arrival as novelties was not through a series of adaptations.

Theoretical rejection of such possibilities is prone to circularity when it appeals to the problematic concept of fitness.[34] Furthermore, experience with invasive species[35] and phenomena such as "transgressive segregation"[36] raise important questions about the requirement for gradualism in the establishment of successful organism-ecosystem relationships. The success of alternative phenotypes that arise through genetic or environmental effects will depend on their settling into or constructing niches[37] in which their types breed true. Competition between variants may accelerate change, but in our view it is not the engine of evolution that it is for the Synthesis. With genetic accommodation (selection after the fact on genetic variation supportive of the new phenotype)[38],

---

[34] Ariew, A., and R. C. Lewontin. 2004. The confusions of fitness. *Brit. J. Phil. Sci.*, 55:347-363.

[35] Carroll, S. P. 2008. Facing change: forms and foundations of contemporary adaptation to biotic invasions. *Mol Ecol*, 17:361-72; Stohlgren, T. J., D. T. Barnett, C. S. Jarnevich, C. Flather, and J. Kartesz. 2008. The myth of plant species saturation. *Ecol Lett*, 11:313-22; discussion 322-6.

[36] Rieseberg, L. H., M. A. Archer, and R. K. Wayne. 1999. Transgressive segregation, adaptation and speciation. *Heredity*, 83 (Pt 4):363-72.

[37] Odling-Smee, F. J., K. N. Laland, and M. W. Feldman. 2003. *Niche construction: the neglected process in evolution*. Princeton University Press, Princeton, N.J.

[38] West-Eberhard, M. J. 2003. *Developmental plasticity and evolution*. Ox-

the development of the novel form could become independent of the inducing environment.

At intermediate stages of evolution where canalization is partial, body plans would not be subject to wholesale change, but many other features would. The origination of novelties like vertebrate limbs,[39] avian feathers,[40] and beetle horns[41] would occur as embellishments within pre-existing body plans based on residual plasticity of the phenotype. The implication of all this is that evolutionary options become more and more constrained over time, and that differences among the higher taxonomic groups are laid down with broad brush strokes before the distinguishing details among sister species are added late in the process. This is the absolute inverse of the Darwinian scenario, where speciation is the primary evolutionary mechanism and (since microevolution is the only means to macroevolution), the most dramatic differences among taxonomic groups (i.e., phyla) take the longest times to become established.

This line of reasoning can be extended even further back in time to help understand the origination of the body plans themselves. At the earliest stages of animal evolution the characteristic organisms were just clusters of cells that aggregated after dividing, or did not separate from one another after cell division. During this period, before any pathways for morphogenesis or pattern formation had evolved,[42] canalization would have been nonexistent. Ancient multicellular forms, the ancestors of the modern animals, would have been completely subject to the physical mechanisms and effects that pertain to all viscoelastic, chemically and mechanically excitable materials. These phenomena include adhesion, which aggregates cells into liquid-like droplets, surface tension, which causes such droplets to round up, phase separation, which leads them to form immiscible layers, molecular diffusion, which causes them to be spatially heterogeneous, and so forth.

---

ford University Press, Oxford; New York.

[39] Newman, S. A., and G. B. Müller. 2005. Origination and innovation in the vertebrate limb skeleton: an epigenetic perspective. *J Exp Zoolog B Mol Dev Evol*, 304:593-609.

[40] Prum, R. O., and A. H. Brush. 2002. The evolutionary origin and diversification of feathers. *Q Rev Biol*, 77:261-95.

[41] Moczek, A. P. 2008. On the origins of novelty in development and evolution. *Bioessays*, 30:432-47.

[42] Newman, S. A., G. Forgacs, and G. B. Müller. 2006. Before programs: The physical origination of multicellular forms. *Int. J. Dev. Biol.*, 50:289-99.

A set of molecules in the single-celled ancestors of the animals[43] were predisposed to assume novel functions in the multicellular state by mobilizing physical effects which had been irrelevant to the scale of the individual cell, in the form of "dynamical patterning modules" (DPMs).[44] In particular, the multicellular state itself likely came into being by one class of ancestral molecules, the cadherins, forming a DPM in conjunction with the physical effect of homophilic adhesion. Because of the limited set of relevant physical effects that apply to matter on this scale, a specific collection of morphological motifs were inevitable in animal systems, constituting a deeply embedded orthogenetic principle in this class of organisms. The recurrent appearance during the course of evolution of interior body cavities, multiple tissue layers, elongated bodies, segments, and appendages do not, therefore, require adaptationist accounts.

This perspective clears up a number of conundrums and apparent paradoxes that unsettle the standard model. Because physics can act immediately to shape and modify form, particularly in the developmental stages of uncanalized organisms, early wide-ranging morphological diversification need not have taken long periods of time. Genetic consolidation likely followed, rather than accompanied, the rapid radiation of body plans referred to as the Cambrian explosion.[45] In addition, while the limited number of physical effects that enter into DPMs are dictated by the laws of nature, the set of molecules involved in these modules is also limited, but this is just a fortuitous matter of their having been the only suitable ones present in the ancestral cells. The suggestion that the animal phyla emerged early and rapidly by means of the DPMs carries the implication that the associated molecules, the "developmental-genetic tool-kit," would be central to the develop-

---

[43] Abedin, M., and N. King. 2008. The premetazoan ancestry of cadherins. *Science*, 319:946-8; King, N., M. J. Westbrook, S. L. Young, A. Kuo, M. Abedin, J. Chapman, S. Fairclough et al. 2008. The genome of the choanoflagellate *Monosiga brevicollis* and the origin of metazoans. *Nature*, 451:783-8.

[44] Newman, S. A., and R. Bhat. 2008. Dynamical patterning modules: physico-genetic determinants of morphological development and evolution. *Phys. Biol.*, 5:15008; Newman, S. A., and R. Bhat. 2009. Dynamical patterning modules: a "pattern language" for development and evolution of multicellular form. *Int J Dev Biol.*, 53:693-705.

[45] Rokas, A., D. Kruger, and S. B. Carroll. 2005. Animal evolution and the molecular signature of radiations compressed in time. *Science*, 310:1933-8; Conway Morris, S. 2006. Darwin's dilemma: the realities of the Cambrian 'explosion'. *Philos Trans R Soc Lond B Biol Sci*, 361:1069-83.

mental pathways of all extant animals.[46] This is in fact the case, but the microevolutionary scenarios of the Synthesis would not have predicted it.

Levins and Lewontin[47] consider the hallmark of Darwin's theory to be a principle of *variation*, whereby individual members of a population differ from each other in some properties and the population evolves by changes in the proportions of the different types. In this process some variant types persist while others disappear, so the nature of the population changes without there being successive changes in the individual members. They contrast this variational principle to a *transformational* one: for Lamarck species changed because individual organisms underwent transformations during their life history. The above description places our model for the early, most dramatic stages of morphological evolution in the transformational category. It includes Darwinism as a limiting case, however: once extensive canalization has set in, individual transformations are less possible and natural selection of small variations becomes the predominant mode of evolutionary change.

3. What, if any, practical and/or social-political and/or moral obligations follow from your work on evolution?

It has been claimed that the Modern Synthesis leaves no place for ultimate meanings. In a sense, our alternative model for multicellular evolution is no different, being entirely materialistic. My own view about ultimate meanings, however, is that the concept represents a category mistake.[48] Meanings are relational properties between entities with motives and goals, and the capacity to learn from experience. As far as science can tell, only living things of sufficient complexity have these qualities.[49] Since life appears to be a relatively late arrival in the universe, meaning is not ultimate, but conditional.

This contention is underdetermined, however. The issue relates

---

[46]Newman, S. A. 2006. The developmental-genetic toolkit and the molecular homology-analogy paradox. *Biological Theory*, 1:12-16.

[47]Levins and Lewontin (1985; see footnote 14) p. 86.

[48]Ryle, G. 1949. *The concept of mind*. Hutchinson's University Library, London, New York.

[49]Ginsburg, S., and E. Jablonka. 2007. The Transition to Experiencing: I. Limited Learning and Limited Experiencing. *Biological Theory*, 2:218-230; II. The Evolution of Associative Learning Based on Feelings. *Biological Theory*, 2:231-243.

to the tenet of "uniformitarianism," the supposition of Darwin, carried over from the geological insights of James Hutton and Charles Lyell, that similar forces drive evolutionary change at all stages.[50] Since the physics and plasticity based evolutionary model described above is only concerned with the transition from single-celled ancestors to multicellular forms, it does not contain a theory of cellular evolution. In trying to understand the origins of cellular life we cannot resort to the Darwinists' confident assertion that it was due to the same set of processes that have purportedly been demonstrated for animals and plants. Our model may be multiscale, but it is also domain specific. So the origin of life, and therefore meaning, remains an open question.

Because of the unknowns involved in early evolution (the avowedly materialistic molecular biologist Francis Crick even speculated that cellular systems may have been manufactured elsewhere than the Earth),[51] and the likelihood, in my view, that the standard model does not adequately account even for multicellular phylogeny, a different approach needs to be taken to public scepticism around the actual facts of evolution. Evolution's defenders usually portray natural selection, or "Darwin's theory of evolution" as its only tenable scientific account. By doing so, they treat calls to open up the discourse around the subject as a threat to science and rationality. This is partly true: evolution's opponents are, in fact, almost uniformly trying to insinuate religious or anti-naturalist ideas into classrooms and textbooks, because they also see natural selection and special creation as the only possible alternatives. As I have argued above, however, multicellular evolution may be better described by non-Darwinian processes that incorporate the physical nature of such systems into the explanatory framework than by a gene-centered theory of non-directional organic change that is often, like the views it is trying to displace, faith-based.[52] As an educator, I have confidence that students and members of the public who are undecided between traditional and scientific narratives will more readily opt for latter to the extent that the science itself is persuasive.[53]

---

[50] Gould, S. J. 1987. Time's arrow, time's cycle: myth and metaphor in the discovery of geological time. Harvard University Press, Cambridge, Mass.

[51] Crick, F. 1981. *Life itself: its origin and nature.* Simon and Schuster, New York.

[52] Lewontin, R. C. 2009. Why Darwin. *The New York Review*, Vol. 56 (May 28), pp. 19-22.

[53] Newman, S. A. 2008. Evolution: the public's problem and the scientists'.

This new view of evolution also has certain implications for biotechnology policies. The prevailing concept of organisms as genetically determined machines whose origination and direction of change over the course of evolution were determined only by criteria of fitness and efficiency, readily lent itself to an engineering paradigm. In particular, we would only need to rewire the appropriate circuits and retune the correct switches (all gene-based) to obtain desired improvements, whether in the realm of crops, or even future humans. If, instead (as I have suggested), present-day organisms have resulted from an interplay between genetic and non-genetic mechanisms whose causal nature has changed over time, tinkering with the genes of the late-stage products is a recipe for disaster, ecologically, socially and individually.[54]

4. What do you see as the most interesting criticism against your position in the biological or philosophical discussion of evolution?

The most comprehensively described developmental system at the level of gene regulatory networks is the sea urchin larva, the long-term project of Eric Davidson and his colleagues.[55] This system has been characterized by this group as computer- or machine-like,[56] and as such, has been taken as a paradigm for the early evolution of multicellular form.[57] Although this system is said to exhibit "plasticity," because changes in its transcription factor-based hardwiring can lead to changes in morphological phenotype,[58] and to evolve in a "nonuniformitarian" fashion, because the hierarchical nature of the gene regulatory networks progressively restricts avenues of further change,[59] the use of these terms is very different from ours. As discussed above, we take plasticity to be the possibility of phenotypic change without genotypic

---

Capitalism Nature Socialism, 19:98-106.

[54] Newman, S. A. 2009. Renatured biology: Getting past postmodernism in the life sciences. In, *Without nature? A new condition for theology*, (eds., D. Albertson and C. King), New York: Fordham Univ. Press. pp. 101-35; 392-403

[55] Davidson, E. H. 2006. *The regulatory genome: gene regulatory networks in development and evolution.* Elsevier/Academic Press, Amsterdam; London.

[56] Istrail, S., S. B. De-Leon, and E. H. Davidson. 2007. The regulatory genome and the computer. *Dev Biol*, 310:187-95.

[57] Erwin, D. H., and E. H. Davidson. 2009. The evolution of hierarchical gene regulatory networks. *Nat Rev Genet*, 10:141-8.

[58] ibid

[59] ibid

change, and a nonuniformitarian evolutionary scenario to imply that the types of formative mechanisms differed at different stages, not simply that the latitude of a single mode of change became more constrained.

Our framework implies that hierarchically-organized, precisely "engineered" developmental systems would have emerged by selection-driven reinforcement and refinement of protean forms originally produced by dynamical patterning modules, i.e., physical mechanisms mobilized by non-transcription-factor gene products.[60] The perspective of Davidson and co-workers, in contrast, implies that the evolution of hardwired machine-like developmental systems from ancestral unicellular organisms proceeded by a series of intermediates that were themselves each strictly programmed by networks of transcription factors.

Davidson and co-workers' view of the evolution of developmental systems is the one most solidly based in the Modern Synthesis and is right now probably the most widely accepted one. It thus represents the perspective that our alternative most starkly challenges.

5. With respect to present and future inquiry, how can the most important problems concerning evolutionary theory be identified and explored?

Organisms, ancient and modern, are embedded and involved in environments that to varying extents form and are formed by them. Living systems have always been physical systems, but their inherent properties and the forces that act on them have changed coordinately with changes in their composition and size. The physical determinants relevant to populations of individual rigid-walled bacterial cells in turbulent media are different from those pertaining to clusters of amoeboid cells. Although cells can communicate in both these situations, the means of communication are different and so are the characteristic modes of collective behaviour. It is these modes and their phenotypic outcomes that provide the raw materials for natural section; they are provided by the dynamics of the systems in question, not by selection drawing on a blank slate.

---

[60] A mathematical model for such "mechanism succession" can be found in Salazar-Ciudad, I., S. A. Newman, and R. Solé. 2001. Phenotypic and dynamical transitions in model genetic networks. I. Emergence of patterns and genotype-phenotype relationships. *Evol. & Develop.*, 3:84-94.

These considerations also apply to scales above and below collectivities of cells. Networks of neurons exhibit specific modes and patterns of excitation but also receive input from the outside world and can therefore respond to, represent, and model it at increasingly higher levels of sophistication.[61] Understanding the evolution of behaviour, cognition, and higher functions such as consciousness and language[62] must take these inherent self-organizational properties into account, but it is not necessary to postulate incremental gene-based scenarios for their construction.

The subcellular world presents similar issues. Since so much is known about the relation of subcellular components to the genes that specify them, the notion has gained hold that the machinery of the cell has been accounted for, at least in principle, by Darwinian scenarios. But in fact this is one of the places where the standard model has proved most vulnerable and unpersuasive. The bacterial flagellum, for example, is a rotating molecular whip consisting of several dozen proteins that functions with amazing precision and efficiency. Envisioning an incremental scenario for the evolution of this macromolecular assemblage is particularly difficult, since most of its proteins are essential to its functioning.[63]

Solving the problem of how the flagellum and a host of other equally complex subcellular "machines" have evolved may require, as we believe understanding multicellular evolution does, an appreciation of the role of plasticity in the generation of novel structures and functions. Although it has long been believed that protein 3D structure is uniquely determined by its primary sequence and thus its genetic encoding, it is now recognized that some proteins can take on distinct alternative structures with distinct functions,[64] and that many other proteins have no intrinsic structure,

---

[61] Hayek, F. A. v. 1952. *The sensory order: an inquiry into the foundations of theoretical psychology*. University of Chicago Press, Chicago; Thelen, E., and L. B. Smith. 1996. A dynamic systems approach to the development of cognition and action, *MIT Press/Bradford Books series in cognitive psychology*, MIT Press, Cambridge, Mass.

[62] Boeckx, C. 2006. *Linguistic minimalism : origins, concepts, methods, and aims*. Oxford University Press, Oxford; New York.

[63] Pallen, M. J., and N. J. Matzke. 2006. From The Origin of Species to the origin of bacterial flagella. *Nat Rev Microbiol*, 4:784-90.

[64] Kimchi-Sarfaty, C., J. M. Oh, I. W. Kim, Z. E. Sauna, A. M. Calcagno, S. V. Ambudkar, and M. M. Gottesman. 2007. A "silent" polymorphism in the MDR1 gene changes substrate specificity. *Science*, 315:525-8.

behaving in an entirely context-dependent fashion.[65] It is thus plausible that the flagellum arose in a series of abrupt steps from earlier multiprotein complexes with very different structural and functional properties.

The new understanding of proteins casts doubt on the assumption of a unique genotype-phenotype relationship at all levels of biological organization.[66] This provides further motivation for expanding evolutionary theory to incude environment-dependent plasticity, self-organization, and saltational change.

---

[65] Gsponer, J., and M. Madan Babu. 2009. The rules of disorder or why disorder rules. *Prog Biophys Mol Biol*, in press; Jeffery, C. J. 2009. Moonlighting proteins–an update. *Mol Biosyst*, 5:345-50.

[66] Newman, S. A., and R. Bhat. 2007. Genes and proteins: dogmas in decline. *J Biosci*, 32:1041-3.

15
Samir Okasha

Professor of Philosophy of Science

Bristol University, UK

---

1. Why were you initially drawn to discussions and research on evolution (or evolutionary aspects of your field)?

I was initially drawn to evolutionary biology while studying for a PhD in philosophy of science at Oxford in the 1990s. This was the heyday of W.D. Hamilton and Richard Dawkins, so evolution was very much 'in the air' at Oxford, and I attended numerous biology seminars during my postgraduate years. At first this was just for interest, as it hadn't occurred to me that it might be possible to integrate evolutionary biology with my work in philosophy of science, which at the time focused on topics such as induction, probability and scientific inference. However, this changed as I became familiar with the burgeoning literature in philosophy of biology, by authors such as Elliott Sober, David Hull, and Alexander Rosenberg. By the time I left Oxford, in 1997, I was certain that I wanted to do work in the philosophy of biology, and I spent most of the next seven years immersing myself in evolutionary theory. It struck me then, and still does today, that a proper understanding of the theoretical side of evolutionary biology is crucially necessary if one wants to understand the philosophical foundations of the field.

In retrospect, I think that I was drawn to evolutionary biology for a number of reasons. One is simply the intrinsic interest of the subject matter. I can vividly remember my sense of intellectual excitement on first reading Maynard Smith's *The Theory of Evolution* and Dawkins' *The Selfish Gene*, as I'm sure countless others can too. But there were also other reasons, relating to philosophy. Darwinian evolution is an extremely powerful idea, with an appealing simplicity and generality, and I have long been convinced that it can shed light on diverse topics, including ones that were traditionally the province of philosophers, e.g. the nature of morality, the nature/nurture debate, the reality of natural kinds, the tension between individual self-interest and group welfare, and others. Of course, this is not a novel suggestion; it is the guiding light behind Dennett's 1995 book *Darwin's Dangerous Idea*, for example. But I think that much work remains to be done in bringing evolutionary biology to bear, in a non-trivial way, on diverse philosophical discussions.

A different source of my interest in evolutionary biology came not from the promise that evolution might illuminate traditional philosophical issues, exciting though that is, but from a realisation that much work in evolutionary theory rests on implicit philosophical assumptions, which can profitably be brought into the open. In my book *Evolution and the Levels of Selection* (Okasha 2006), I tried to show how the debate in evolutionary biology over units and levels of selection involves a subtle intertwining of empirical and philosophical (or conceptual) questions, often not sharply separated. For example, many disagreements about the levels of selection can be traced, in part, to the protagonists' differing assumptions about causality, explanation and reductionism, and are thus not purely empirical. As I see it, a major task for philosophy of science is to scrutinize these implicit assumptions that we find at work in scientific debates, thus providing conceptual clarification. (Philosophers of physics have long performed a similar task in relation to debates in physics.) Typically, this sort of philosophical project requires doing foundational work in the science itself.

A final source of my interest in evolution, and one which motivates my current research project, is the striking parallelism between evolutionary theory and economics. (My undergraduate degree was in economics and philosophy, so I was struck by the evolution/economics connection as soon as I began to study biology.) This parallelism, which is both formal and thematic, arises because maximisation or optimization is central to both areas. One way to appreciate this is to see that the notion of fitness in evolutionary biology plays a similar role to the notion of utility in economics and decision theory; one is the quantity maximised by the evolutionary process, the other by the rational agent. This observation is not particularly novel; it is the key theme of a 1995 book by Brian Skyrms (Skyrms 1995), it is widely appreciated by evolutionary game theorists, and is also related to some recent work by Alan Grafen, discussed in section 5 below. However, much work remains to be done to fully understand the fitness/utility connection; this is the subject of my current research project, jointly with the game theorist Ken Binmore here at Bristol.

2. What does your work reveal about biological evolution (or evolutionary aspects of your field) that other academics, citizens, philosophers or biologists typically fail to appreciate?

My own work aims to straddle philosophy of science and evolutionary theory, so it doesn't really involve making new discoveries about evolution. Rather, the aim is to clarify concepts, explore connections between theories, and scrutinize the implicit philosophical assumptions that often underpin the biological work. I think that my work on levels of selection highlights the pervasive role that such assumptions can play in science, something that is not always recognised. Relatedly, I have tried to show how empirical and conceptual questions in evolutionary biology often become inter-twined, and to develop some methods for separating them. Again, I think that this is not always fully recognised.

I will give two examples to illustrate this. In the debate over levels of selection, part of the problem is that different authors invoke, usually implicitly, different criteria for what counts as a 'level' in the biological hierarchy, and/or different criteria for when selection can be said to occur at a given level. One standard idea, enshrined in the 'Price equation' approach to multi-level selection, and suggested by Lewontin's famous 1970 paper *The Units of Selection* is that selection occurs at any level where there is character-fitness covariance. However this idea fails to take account of what I call 'cross-level byproducts', i.e. that a given character-fitness covariance might be an indirect side-effect of the causal action of natural selection at a *different* level, higher or lower. Many of the ideas in the literature about how to identify the 'true' level(s) of selection, are in effect attempts to say when a given covariance is a cross-level byproduct. This is true, for example, of the appeal to 'emergent characters', often made in this context. When the levels of selection issue is framed in this way, a point of contact is established with the more general issue of causation in hierarchically structured systems, and the related philosophical literature on reductionism. This enables a fresh, and unifying, perspective on the original biological problem.

Another example comes from the literature on 'major evolutionary transitions', which occur when a number of free-living units combine themselves into a larger whole, sacrificing their individuality and giving rise to a new level of biological organization (cf. Michod 1999, Maynard Smith and Szathmary 1995). (Think for example of the transitions from prokaryotic to eukaryotic cells, from single-celled to multi-celled eukaryotes, or from solitary to colonial organisms.) Such transitions involve a form of multi-level selection – for selection can act on the coalescing individuals themselves, and also on the groups or proto-organisms that they are in the process of forming. This point is quite widely acknowledged. However, there are two quite different types of multi-level selection – sometimes called 'MLS1' and 'MLS2' (cf. Damuth and Heisler 1988); in the latter, the groups themselves engage in a form of reproduction, begetting offspring groups, while in the former, the groups are merely part of the environment of the individuals, so do not themselves reproduce. This distinction corresponds roughly to the difference between 'group selection' and 'species selection', in the early levels of selection literature. In my 2006 book, I try to show that in a typical evolutionary transition, both types of multi-level selection must occur, but at different temporal stages. Early on in a transition, before the higher-level groups have formed, MLS1 is the relevant sort of multi-level selection, but in later transitional stages MLS2 becomes applicable. So what appeared to be a mere ambiguity in the notion of multi-level selection is in fact quite significant, for it helps us understand the sort of fitness re-organisation that occurs during evolutionary transitions in individuality.

More generally, within the levels of selection debate, I have tried to integrate aspects of the multi-level or hierarchical selection approach with the more reductionistic 'gene's eye' approach to evolution, arguing that these approaches are in fact complimentary, not antithetical. (This is a generalization of the argument, once regarded as heretical but by now quite widely accepted, which says that kin and group selection are ultimately equivalent ways of viewing social evolution.) The fact that the genic and multi-level approaches are not really in opposition is a lesson that, although understood by specialists, isn't fully appreciated in the wider intellectual community.

3. What, if any, practical and/or social-political and/or moral obligations follow from your work on evolution?

None really. My work's a bit too abstract to have either practical or moral/socio-political implications. Moreover, I think that one must be rather cautious in trying to extract socio-political implications from evolutionary ideas. It's not impossible that there might be such implications. For example, evolutionary psychologists might conceivably make discoveries about human psychology which could inform social and economic policy. Or evolutionary game theorists might produce findings which imply that certain sorts of social reform, or certain ways of trying to organise society, will not work. But I don't think this is actually very likely. And the history of attempts to draw social and political implications from evolutionary theories, from Spencer's social Darwinism to Fisher's eugenics to the modern debate on the heritability of IQ, should teach us to tread with caution here.

Although my own work on evolution doesn't have direct practical implications, that of others may well do. The burgeoning field of 'evolutionary medicine' might be an example. Or conceivably, theories of evolutionary dynamics could be put to use in devising methods for controlling the spread of viruses, or for designing conservation policies. But these are *practical* implications, not moral or socio-political ones. About the latter, I'm rather sceptical, as I'm a believer in the old-fashioned Humean view that you can't get an 'ought' from an 'is'.

I do think that evolutionary biology can tell us a lot about morality, but this is a rather different matter. There is a lot of very interesting work on the evolution of human moral sentiments, e.g. our sense of 'fairness', on the evolution of the social contract, and of course on the evolution of altruistic and pro-social behaviour. I think that traditional moral philosophers would do well to pay attention to all this work; though of course they will argue that it doesn't address the 'normative' questions that they are mostly interested in. That may be so, but an understanding of how our moral psychology evolved, of the sort that evolutionary biology promises to provide, must surely be of interest to anyone concerned with ethical questions. The tendency of traditional ethicists to demarcate their field of study so narrowly that these evolutionary ideas are judged irrelevant to it, strikes me as a pity, and unfaithful to the history of their subject. Had they been alive today, Aristotle and Hume would surely have been fascinated by the prospects of an evolutionary approach to the study of morality.

4. What do you see as the most interesting criticism against your position in the biological or philosophical discussion of evolution?

Well, I haven't really tried to stake out a 'position' as such; in fact, I think that many philosophers, including philosophers of science, have spent too much time developing overarching 'positions', as opposed to engaging in careful, piecemeal analysis of local issues. Witness the embarrassing proliferation of 'isms' in the contemporary philosophical literature.

One assumption that underpins a lot of my work, and that could certainly be criticised, is that scientific theorising often rests on philosophical presuppositions, and thus that there is value in attempting a philosophical scrutiny of contemporary science, including evolutionary biology. I suspect that many biologists would not sign up to this idea, or at least would argue that I have overstated the case.

A good example of this comes from my work on levels of selection, some of which I described above. A major part of my project was to explore the links between different ways of conceptualising the levels of selection problem, and between ways of mathematically modelling selection in a hierarchical setting. My aim in doing this was to clarify concepts, to provide a general schema for thinking about selection at multiple levels, to develop a taxonomy of selection processes, and to address philosophical issues raised by the levels of selection discussion. However, a critic might argue that at root, the issues I was addressing were not conceptual but semantic. For example, my discussion of different concepts of multi-level selection, which I had hoped would help clear up some confusions in the biological literature, could be re-described, by a harsh critic, as merely a discussion of different ways of using the expression 'multi-level selection'. I would of course resist such a re-description, but it is not always easy to sharply demarcate conceptual from semantic issues. So the critic's objection is not easy to refute.

A criticism of this sort – and in my view an unfair one – was made by John Maynard Smith in his review of Sober and Wilson's book *Unto Others*. In that book, Sober and Wilson devote considerable space to arguing that models of both kin selection and of frequency-dependent selection typically involve a component of group selection, and that this is why they permit altruistic behaviours to spread. Thus it is quite wrong to think that modern social evolution theory has dispensed with group or multi-level selection, Sober and Wilson claim, despite what is often thought. In response, Maynard Smith accused Sober and Wilson of simply having re-defined 'group', so that any interacting organisms automatically constitute a group, however few they are and however fleeting the interaction. But Maynard Smith's attempt to dismiss the issue as semantic is unhelpful, I think. It is true that Sober and Wilson use the expression 'group' in a broad sense, as they admit. But their point is that there is a deep structural similarity between kin, frequency-dependent and group selection, which is often obscured; their broad definition of 'group' is meant to bring out this similarity. So the issue is not just semantic, but is about the theoretical similarities between different evolutionary mechanisms, and about whether they should be co-classified. Something similar applies to my work, I think.

Turning to my current research project, on the thematic connections between evolutionary theory and economics, one potential criticism is that these connections are actually less interesting than I make them out to be. I argue that the centrality of the notion of optimization in both areas, and the close formal similarities between the theories in both, are theoretically interesting and worthy of examination. But a critic might argue that these similarities are merely co-incidences. Take for example the case of game theory. Like many others, I have long been impressed by the fact that many game-theoretical models can either be interpreted 'classically', i.e. as a model of the deliberation of rational agents, or 'evolutionarily', i.e. as a model of evolutionary adjustment in a population. This fact is deeply interesting in my view, despite its familiarity, and cries out for philosophical analysis. But a critic might dispute this, on the grounds that formal similarities between models in different areas of science often crop up, and don't always show us something deep. I am confident that such an objection would be misplaced, in the case at hand, but I would be hard pressed to offer a definitive argument to this effect.

5. With respect to present and future inquiry, how can the most important problems concerning evolutionary theory (or evolutionary aspects of your field) be identified and explored?

I wouldn't like to say what the most important problems in evolutionary theory are; however, I will mention a few developments that strike me as particularly interesting. One is Alan Grafen's 'Formal Darwinism' project, and in particular his aim of establishing firm links between evolutionary dynamics and optimization notions in biology, in a formally precise way. Grafen's project, though abstract, address fundamental issues in Darwinian theory, and I think that philosophers of biology would do well to pay close attention to it.

Another interesting problem concerns the integration of traditional evolutionary theory with the field of evolutionary developmental biology. Evo-devo is an exciting area at the moment, and has produced striking new empirical discoveries, but there has been relatively little work by evolutionary theorists that explicitly takes account of these discoveries. Some proponents of evo-devo see their work as marking a major break with neo-Darwinism, while others dispute this. One way of addressing this issue might be to ask how traditional evolutionary models and theories need modifying, if at all, to incorporate a developmental perspective. For example, how much does classical population genetics need modifying to take account of the fact, discovered by evo-devo, that much evolutionary change involves change in regulatory, rather than coding, regions of the genome?

On a different matter, it would be interesting to see evolutionary theorists extend the neo-Darwinian paradigm to a broader class of species than it is usually applied to, including ones where the notions of 'individual' and 'organism' are hard to apply, e.g. many clonal species. In his seminal book *The Evolution of Individuality*, Leo Buss argued that neo-Darwinian theorists had unwittingly restricted the scope of their theory, by incorporating into its foundations certain assumptions that while true of metazoans, were not true across the living world. These included: the non-inheritance of acquired characters, sequestration of the germ line, the evolutionary insignificance of within-organism genetic change, and others. Buss argued that these assumptions, which are not true of many non-metazoan taxa, needed to be relaxed in order for a fully general evolutionary theory to be developed. I think that Buss was on to a deep point, and posed a challenge for evolutionary theorists that has yet to be met.

Finally, closer to my own current research interests, I think there is interesting work to be done in integrating evolutionary theory with theories from the social sciences, particularly economics. Early attempts to do this, e.g. by the sociobiologists, were not particularly successful, and were often perceived as hostile takeover bids, by the social scientists. But much of the recent work on cultural evolution, and on evolutionary economics, strikes me as far more promising, and capable of genuinely enriching social science, rather than trying to take it over.

Within the field of philosophy of biology, rather than evolutionary theory itself, there is plenty of promising work to be done. However, I have been slightly disappointed that much recent work in the area hasn't really achieved much contact with the biology itself. To some extent this is inevitable, as the issues that excite philosophers of science are not always those that excite the scientists themselves. But I think it's crucial that the philosophy of biology literature does not become self-contained, and that genuine contact with the biology is established and maintained.

References

Maynard Smith, J. (1966) *The Theory of Evolution*, 2nd edition, London: Penguin.

Dawkins, R. (1976) *The Selfish Gene*, Oxford: Oxford University Press.

Dennett, D. (1995) *Darwin's Dangerous Idea*, London: Penguin.

Okasha. S. (2006) *Evolution and the Levels of Selection*, Oxford: Oxford University Press.

Skyrms, B. (1995) *Evolution of the Social Contract*, Cambridge: Cambridge University Press.

Lewontin, R. C. (1970) 'The Units of Selection', *Annual Review of Ecology and Systematics*, 1, 1-18.

Michod, R. (1999) *Darwinian Dynamics: Evolutionary Transitions in Fitness and Individuality*, Princeton: Princeton University Press.

Maynard Smith, J. and Szathmary, E. (1995) *The Major Transitions in Evolution*, Oxford: Oxford University Press.

Damuth, J. and Heisler, I. L. (1988) 'Alternative Formulations of Multi-level Selection', *Biology and Philosophy* 3, 407-30.

Maynard Smith, J. (1998) 'The Origin of Altruism', *Nature* 393, 639-40.

Sober, E. and Wilson, D. S. (1998) *Unto Others: The Evolution and Psychology of Unselfish Behaviour*, Cambridge MA: Princeton University Press.

Buss, L. W. (1987) *The Evolution of Individuality*, Princeton: Princeton University Press.

Grafen, A. (2007) 'The formal Darwinism project: a mid-term report', *Journal of Evolutionary Biology* 20, 1243-1254.

Grafen, A. (2008) 'The simplest formal argument for fitness optimisation'. *Journal of Genetics*, 87, 421-433.

16
Susan Oyama

Emerita Professor of Psychology
John Jay College
CUNY Graduate Center, USA

---

1. Why were you initially drawn to discussions and research on evolution (or evolutionary aspects of your field)?

My work in what is now called Developmental Systems Theory (DST) actually began not with evolution *per se* but with the nature-nurture opposition. It was only after lengthy explorations of the issues of developmental regulation and control, first in language, my doctoral specialty, then in psychology and biology more broadly, that I saw that conceptions of development and evolution were so intertwined that serious work on the nature-nurture problem required engaging with evolutionary theory as well. By "serious work" I do not mean placing phenomena on that continuum that is said to stretch from biological hyperdeterminism to a la-la land of giddily unmoored "social constructivism," neither pole recognizable except as a caricature. Rather, I mean evaluating its structure, assumptions and stock inferences.

As I became progressively more involved with evolutionists and their writings, as well as the (mis)uses of their work by people in other disciplines, it became obvious that evolutionary and developmental studies were locked in a mutually supportive relation: theoretical codependency, if you will, to borrow from pop-psychology. Beyond this reciprocal reinforcement, the two literatures exhibit oddly parallel conceptual structures, in which internal and external causation are contrasted, primary importance typically being awarded to insides by developmentalists (especially embryologists) and outsides by evolutionists (Gray, 1987, 1989; Oyama 1992/2000a).

Even though some developmentalists had realized for decades that nature-nurture distinctions were riddled with ambiguities and incoherencies, and even though it was for a period *de rigueur* to dismiss them as passé, an ever-more-numerous and influential flock of "evolutionary approaches" almost universally mandated some form of that dualism: between instincts shaped by natural selection and habits formed by learning, innate and acquired traits, closed and open programs. Such is the prestige of evolutionary theory, in fact, augmented by astounding technological advances in genetics and molecular biology, that it is quite common to find blandly unapologetic references to innateness and instinct in the very fields that had become so wary of them in the second half of the last century. Evolutionary psychologists now proliferate cognitive modules and innate psychological mechanisms with the same abandon that instincts were inventoried a century earlier.[1]

Meanwhile, students of evolution (Gould & Lewontin, 1979; Maynard Smith, Burian, Kauffman, Alberch, Campbell, Goodwin, Lande, Raup, & Wolpert,1985) argued about internal developmental (or physical) constraints on a natural selection construed as the shaping of species by the environment.[2] Recurring controversies over Lamarckian and Darwinian evolution tend similarly to be couched in terms of internal and external forces.

Not surprisingly, key articulations between ontogeny and phylogeny typically implicate nature-nurture dualisms; these nodes include hereditary transmission, genetic assimilation and the Baldwin effect (Weber & Depew, 2003), and, of course, genetic information, the coded instructions, passed on in reproduction, that supposedly make the next generation. Undoing the nature-nurture knot thus entails un- and re-doing parts of the standard theories themselves, reworking them in a systems framework to address both the ontogenetic and phylogenetic timescales without feeding the reverberating circuit that keeps unworkable dichotomies in

---

[1] The situation is somewhat muddled by certain efforts in what Poirier, Faucher, and Lachapelle (2005) call GOFEP, or Good Old-Fashioned Evolutionary Psychology, to improve its conceptual underpinnings. (The acronymn captures the intimate relations between GOFEP's cognitivism and Good Old-Fashioned Artificial Intelligence.) Tooby and Cosmides' (1992) repeated invocations of the "developmentally relevant environment," for instance, can sound unnervingly developmental systems-like at times; but their attempt to finesse their nativism sits awkwardly in the nature-nurture framework that predominates in evolutionary psychology. Occasionally overlapping terminology aside, the thrust of GOFEP is antithetical to DST, resting as it does on inherited genetic *representations*.

[2] See Amundson's (1994) distinction between constraints on form and constraints on adaptation.

play. Trying to criticize nativism while accepting prevailing treatments of evolution is like fighting with the proverbial one arm tied behind you: however determined and passionate you are, you're badly handicapped. And trying to deal with certain problems in evolutionary theory with a superannuated theory of development is like clambering into the ring with your *other* arm bound.

DST gives us an evolutionary process constituted by a succession of overlapping developmental systems. These vary in many ways but can also be quite transgenerationally stable, and both variations and constancies are important in understanding their evolution (*contra* Griffiths & Gray, 1994, 2001). Consisting of organisms (including their DNA) and the changing complex of environmental factors that affect their development, developmental systems are extended, heterogeneous, and interconnected, such that the developing organism changes its developmentally relevant surround as it is changed by it. It is in the workings of these systems that the subject matters of developmentalists and evolutionists come together, and they are workings that replace inconsistent nature-nurture contrasts with organisms whose phenotypic natures are the continuing products of developmental nurture.

Having been drawn to evolutionary biology by the mutual implication of its processes with developmental ones, then, I lingered because I was intrigued by the relationships I found there. So I'm the guest who never went home, because it was clear the party wasn't over yet.

2. What does your work reveal about biological evolution (or evolutionary aspects of your field) that other academics, citizens, philosophers or biologists typically fail to appreciate?

Conventionally, development is supposed to be goal-directed, its "programmed" predictability explained by internal controls. Some room may be left for effects from the outside, but these are considered evolutionarily irrelevant because evolution has come to be defined in terms of changing genetic frequencies. Thus the actual phenotype, the organism, becomes an evolutionary dead end, interesting only for the genes that it contributes to the next round (including some indirect effects, on kin, perhaps). Evolution, by contrast, is said to be undirected, unpredictable, and largely externally driven, by the environmental contingencies of natural selection. Organisms must solve the "problems" set for them by the environment (Lewontin, 1983/2001).

All this calls for separate causes, some more fundamental than others (driving forces contrasting with mere perturbations or constraints, basic forms with accidental details, and so on), in what I've called "causal privileging." Such privileging maintains the causal dualism of the conceptual structures just described. Replace accounts that prejudge causal importance with ones that accommodate influences from disparate sources, and what one gets are complexes of heterogeneous, interrelated causes acting on both ontogenetic and phylogenetic time scales: in short, developmental systems whose fullness allows investigatory focus on one or a few factors but prevents the rest from disappearing and thus captures their interdependencies.

DST argues that organisms come into being in such systems, which consist of factors both inside the skin and outside it. Because the effects of any factor depend on the rest of the system, many of the usual distinctions between internal (genetic) and external (environmental) formation become meaningless. And because there is no control center in the nucleus directing all these processes (a fact arguably revealed by the actual *findings* of molecular biology, despite the genocentric *language* in which the findings are generally couched: Jablonka & Lamb, 1995, 2005; Keller, 2000; Moss, 2003; Neumann-Held, 1999) the usual contrast between preprogrammed innateness and chance environmental effects fails as well, at least if it is meant to say anything about developmental causation. All features, including the species-typical, need an adequate environment to develop, and the effects of any developmental factor depend on (are contingent on) the rest of the system. It thus becomes impossible to divide the organism into parts that are attributable to an encoded internal nature and parts that come from external nurture, or to invoke some kind of gene-environment continuum.

It is still possible to distinguish common features from uncommon ones, characteristics that are present at birth from those that appear later, traits shared by evolutionary ancestors from those unique to a species, and so on, but none of these requires the developmental dualism of the nature-nurture dichotomy. In fact, it becomes possible to disambiguate the host of questions that travel under that rubric, so that they can actually be investigated.

Evolution becomes, in a definition I favor, change in the constitution and distribution of these developmental systems: organism-environment complexes reconstructed in each generation from resources provided by previous system iterations or arranged, sought, altered, or brought into being by their ongoing functioning.

In keeping with what I said for Question 1, then, what is "revealed" about evolution by work on developmental systems is that development and evolution are not sharply distinguished *as processes*, one internally driven and the other involving "shaping" from the outside; nor do they make different parts of the organism. Another way of saying this is that taking an evolutionary perspective (on development, on human relations, etc.) need not commit one to the usual impedimenta of a "biological approach"— instincts, one-way flow of genetic information, massively modular, hard-wired cognitive "architecture." Instead, it means paying attention to the ways that successive life cycles of organisms are prepared (or not) by previous ones, and in turn prepare (or not) those to follow. This includes, but is not limited to, the reconstruction of DNA sequences during ontogeny and as part of the provisioning of the next generation. As variant systems are reconstructed with varying frequency, populations change over long timespans. Notice that factors external to the organism are included in this account, as aspects of developmental systems whose effects on an organism in turn affect its chances of helping to generate additional life cycles.

In an important sense, it is *the empirical findings of developmental and evolutionary studies themselves* that "reveal" that development and evolution are a matter of viewing developmental systems on two timescales. I put *reveal* in scare quotes, however, for I believe that science is not made of bare facts that conveniently display the truth for us (Oyama, 2000a, ch. 8, 2006b). If that were the case, there would be no need for argument of the sort described here. These discussions occur precisely because conceptual framing *matters*. It matters within science, shaping questions, assumptions, explanations, and thus, the connections that are drawn with other work, past and present. And these in turn have a profound influence on the charting (and funding) of future efforts, as well as on the applications of research findings and their scientific and popular uptake.

What developmental systems thinking offers is a reworked conceptual framework that relates ontogenies and phylogenies in a different way. It therefore has the potential to affect not just the practice of biologists but our very understanding of what it is that biologists do—what biology *is:* not the science that provides the innate "base" that preexists the experience of individual organisms, but a reconfigured discipline that sees life cycles in their multi-leveled complexity. But what is offered is not necessarily accepted, and therefore not necessarily experienced as *revealed*.

3. What, if any, practical and/or social-political and/or moral obligations follow from your work on evolution?

I actually think it's a mistake to look to conceptual frameworks like this for moral or political obligations. If there's a "lesson" to be taken from DST it is likely to look, initially at least, like a largely negative one, meant precisely to counter generations of attempts to draw social or moral dicta from biology.

People's reasons for seeking guidance from biology, especially evolutionary biology, fall roughly into two categories. The first involves what I have called *incidence*: is the feature hard to prevent or to alter; is it universal; does it appear early in development, etc. These are in principle empirical: theoretically one could gather evidence on them, though practical, ethical, or technical considerations might make this difficult or impossible in fact. Note also that they are associated with the "nature" side of the age-old dichotomy, supposedly formed and driven from the inside and minimally subject to the conditions of development. In a sense they are *preformed*, insofar as they are said to be encoded in the genes, and thus not to have real developmental histories, only evolutionary ones. Thus people have looked to biology to learn what conjugal or economic relations are natural (and therefore desirable, perhaps) and conversely, which ones are unnatural (and therefore to be punished, treated by experts, medicated, or otherwise managed).

The second sort of reason people look to biology for direction is more subtle. Harder to express or to tie to actual observations, these are nevertheless part of the hunt for authoritative "biological bases." I've called them questions of *essence*, to contrast with *incidence* (Oyama, 2002). They rest on an assumption that whether or not it is actually manifested, there exists some internal truth about the being in question. Although incidence and essence are seldom distinguished, and indeed are often assumed to go together, they are not the same. Because essences are unseen, furthermore, they are rather trickier to deal with.

Received wisdom holds that modern evolutionary theory did away with typology and essentialism, but this is a partial truth. A naïve biological determinist view may be refuted, for instance, by pointing to modification by learning, or preventative or meliorative measures (affecting *incidence*), but these do not disturb the conviction that there is still an underlying *something* that carries the enduring nature (*essence*). "You can straighten a worm," Mark Twain reportedly said, "but the crook is in him and only waiting"[3].

When people ask whether humans are naturally (biologically, innately) selfish, aggressive, altruistic, cooperative, nature-loving, heterosexual, etc., they are not necessarily wondering whether these characteristics are fixed and inevitable. They may instead want to know what the true nature of the organism is *despite* the appearance of its phenotype. I use *appearance* advisedly, for the genotype has traditionally been identified with essences, while the phenotype has been mere appearance,[4] the contingent manifestation of a nature that may sometimes be obscured but is always there. Peter Godfrey-Smith (2000) has remarked on DST's "extreme anti-preformationism," but I would give comparable attention to its anti-essentialism; in the present context they are equally important in blocking ill-considered moral pronouncements.

Connecting incidence with essence is the idea that genes carry *representations* of what the organism should become, however it may actually turn out (see Note 1). This helps explain the insistence with which the genes are spoken of in linguistic terms, both in the technical language of molecular and evolutionary biology (translation, transcription, coding, editing) and in more openly metaphorical usage (DNA as the language in which God wrote the book of life, chromosomes as library).

This connection to the Word is not accidental, either historically or thematically: like divine *Logos*, what I call *Biologos* is thought by many to reveal to us what is good–for individuals, for societies, for the species as a whole. At the same time, if we treat genes as agents that can trump the will of the beings in which they reside (don't blame me, blame my DNA), we may think they'll tell us what we can be held responsible for. Certainly the discoveries of molecular biology are not necessary to make these moves; each era's science has found interpreters eager to turn the findings of the times to morals and politics, with sometimes appalling consequences. No such effort has ever, to my knowledge, given a convincing justification for grounding its imperatives in biology.

---

[3] www.twainquotes.com/Reform.html

[4] My audiences still confirm that their early courses in biology defined the *genotype* as the information in the DNA, while characterizing the *phenotype* as the *appearance* of that information. This is provocatively consilient with many religious distinctions between immortal, immaterial souls and transient physical bodies. See Nelkin & Lindee, 1995; Oyama, 2000b, 2009a.

So no, evolutionary biology doesn't tell you what's good, or what you should do. It doesn't tell you how men and women should treat each other, or whether an individual is culpable for some act. This negative role is an extremely important one, despite the negative overtones of *negative*.

Some have read lessons of a different sort in the dynamic systems adverted to in DST: the complex interdependencies of such systems might seem to suggest that nature is somehow *nicer* than it appears in other theories, that there exists a natural harmony, perhaps, or that beings are naturally cooperative rather than competitive (Oyama, 2000a, ch. 8, 11). But exploitation and competition are bred by causal interdependency as much as cooperation is, and when a system "regulates" itself or maintains "balance," these are aggregate-level descriptors saying little of the fate of the component parts. Think of the plants and animals that disappear in turn when ecological succession restores a burned forest. This difference between levels of description may be clearer when *we* are the "parts." Consider classical economics' invisible hand, then think of the families, workers and companies that are so efficiently "regulated."

Biology has often allowed, and sometimes encouraged, unfortunate policy. The solution, though, is not to find a better biology, to supply more trustworthy moral guidance. DST's reworked perspective doesn't produce such devoutly wished-for answers, then, but it does push us to be clear about the questions themselves. It also blocks the attempt to answer them by finding the right balance of internal and external control (mostly innate? partly environmental?), rather than directly addressing the complexities of real life and the uncertainties and possibilities that go with them.

I suggest that the search for the authoritative solution, the timeless truth that will dispel doubt and tell us what must be and what should be, ends up distracting us from living our lives, with all the diversity, shared interests, and sometimes irreconcilable differences that living together entails. But this suggestion hardly "follows" from my views on evolution. Nor is this systems-based picture of the world a value-free rendering of Nature. Whatever our theoretical preferences, though, it seems fair to say that from now on we will have to attend much more closely than we have until now (practical obligation?), not just to a larger range of consequences of our actions, but to consequences of those consequences, as well as to the often startling links that appear among such cascades of change. By the same token we are increasingly forced to take account of the multiplicity of material requirements for any kind of life to persist. It may well be prudent to stockpile seeds for an unknown future, for instance, but although a seed is more than naked DNA strands, there is a danger of fixating on its encapsulated potency and forgetting that without a host of other requirements, changing over time and indefinitely extended in space (in short, a developmental system), those hoarded seeds will not help us much.

Systems approaches of various types have been around a long time. Like any other scientific production, they emerge from specific times, places, and social situations. Perhaps it is happenstance that theories emphasizing context-dependence, ecological embeddedness, systemic integration and interconnectedness appear more numerous at a time when increasingly grave environmental crises are (finally) filtering into general awareness and when the global reach of our economic and political relations is becoming evident in ways both enticing (all those new markets!!) and frightening (they're taking our jobs!). This is neither a question of science following society nor vice versa, as though they were different realms. Rather, we are all in this together.[5]

4. What do you see as the most interesting criticism against your position in the biological or philosophical discussion of evolution?

I have been engaged by a loose set of criticisms, some of which are quite recent, circling around the treatment of the organism. Some seem more a *worry* than a complaint, over DST's challenge to the usual enclosure of developmental processes within the skin. Descriptions of the constant traffic across the boundaries of the emerging entity may make some feel that the organism itself is dissolving.[6] A bit more pointed is Evelyn Fox Keller's (2001) claim that DST tends to "elide" the body, treating it as mere environment to the genes. From a different vantage point, Thomas Pradeu (submitted) has argued that the developmental systems approach, at least in some of its incarnations (he cites especially Griffiths & Gray, 1994, 2001) sacrifices organismic identity. In the aftermath

---

[5] For an astute discussion of the politics of ecology, see Haila, 1997.

[6] Part of this response may be due to a surprisingly common slip made by readers who, despite frequent and explicit statements to the contrary, take the developmental system to be coextensive with, synonymous with, the organism.

of a recent conference, moreover, it was argued that DST needs not only to define the organism, but explicitly to theorize it as such (Life & Mind Seminars, 2008).

I distill from these only partially-converging concerns a few oversimplified questions: Is there any sense in which my views marginalize, dissolve, or ignore the organism? Does my approach to evolution require a formal definition or a theory of the organism? Following on this second, one could ask, is there already a theory of the organism that is both consistent with DST and capable of being integrated with it so as to remedy the alleged lack?

To the first, I think I have an answer. To the second and third, I am not sure, but am intrigued by the challenges, partly because they intersect with issues that have long concerned me.

The easier one first, then. Although I can see why some might feel that the organism is disappearing before their eyes when they read developmental systems-style writings, the fear is misplaced. It is possible, however, that my colleagues Russell Gray and Paul Griffiths, the principals in Pradeu's critique, would disagree. In their (2001), for instance, they deny the possibility of distinguishing between organism and environment (hence my use, above, of *I* and *my*).

Take the definition of evolution I gave in my response to Question 2: change in the constitution and distribution of developmental systems. True, organisms are not explicitly mentioned. But they are present in those systems, which include them and are partly organized by them. More to the point, a developmental system is defined by reference to its organism, and when we are focusing on that level, it is the organism, its development, its activities and its changing nature (the phenotype in transition), that is the object of explanation.

Keller's (2001) charge that the body is treated as mere environment to genes misses DST's multi-leveled approach. This oversight in turn may derive from her own focus on the cell as the "unit of development" (see my 2006a for a fuller response). The point is that in a cell, constituents are "environment" *to each other*: they influence each other's operation and effects. Stretches of DNA are part of the developmental context for other stretches, as one organelle is for another. All these complexes are within the body with which Keller was concerned (she took the unicellular organism as the prototypical body), but far from being slighted, that body is what simultaneously organizes its constituents and is made/organized by them, along with its interactions with influences external to it. It is precisely *its constitution and operation we are talking about*. Similarly, a multicellular organismic body could be seen as environment to its parts, but it is itself in an environment, much of which is developmentally relevant and thus part of its developmental system; there is a tradition of speaking of organisms' *internal* environments as well.

By replacing the language of information with the concreteness of three-dimensional molecules, animals, and plants, furthermore, by fleshing out the abstract language of gene flow with actual interactions in time and space, DST could be said to recover bodies and processes, things and events from the disembodiment that is a common consequence of fervent infophilia (Hayles, 1999; Kay, 2000).

On the sharper challenge, to theorize the organism, as I said, things are not so clear to me. Some participants in the Life & Mind exchange cited above have suggested that the theory of autopoiesis associated with Humberto Maturana, Francisco Varela (Maturana & Varela, 1987; Varela, 1979), and more recently, Evan Thompson (2007), has the focus on organismic (or cellular) unity and identity that DST lacks.

Yet it is important to take into account what a scholar is aiming for. There is a difference between setting out to promulgate a theory of life or of the organism[7], say, and wanting to understand how organisms develop and evolve. Taking the last as my topic, I have proceeded in a pragmatic way, working with whatever tacit notions of the organism appeared to be in play, while directing theoretical energy at the larger system that includes it and the smaller ones within it. There is always a possibility of revising those notions, and elsewhere (2006a) I have hinted at how difficult I think it would be to attempt a rigorous definition. In fact, though, I have never (until, arguably, now) been involved in an exchange in which this issue actually came up, much less functioned as a conversation-stopper, such that it was deemed necessary to pause and insert a definition before continuing.

It may be that I find these efforts interesting in part because I myself have long been concerned with issues of theoretical compatibility, rapprochement, and integration, and have periodically wondered when it is desirable to seek fine-grained agreement and when we (collectively) would benefit more from a heterogeneous field of possibilities (Oyama, 2009b). During a certain period, I

---

[7] The focus of autopoiesis, like Keller's, was actually on the cell.

was often on the same conference panels with people like Varela, Stuart Kauffman, and Brian Goodwin. In many ways they were obvious allies, and questions about the relations between my views and theirs came up constantly. Because their approaches felt to me somewhat internalist, furthermore, and because a large part of DST's conceptual struggle has been to distinguish its organism-environment systems from organisms—or environments–themselves (see Note 5), much attention has been trained on the treatment of boundaries and causal relationships across those boundaries. Recently, in fact, Varela's collaborator, Evan Thompson (2007, esp. ch. 7), has attempted to integrate the notions of self-defining autopoietic systems and developmental systems into a unified account of evolution.

I still believe that remembering what people are trying to do can help us accept the inevitable differences in strategy and emphasis that come from pursuing different projects. And I am still uncertain just how to answer the demand that I define the organism, the individual, or life itself. Temperamentally I am not inclined to such gestures, but now that the challenge has my full attention, I'll wait to see what happens next.

5. With respect to present and future inquiry, how can the most important problems concerning evolutionary theory (or evolutionary aspects of your field) be identified and explored?

In current evolutionary studies there appears to be a pervasive impulse to open up and out, to press against the achievements and strictures of the last century's Evolutionary Synthesis. This creates a need for rigorous accounts of "newly" apprehended multiplicity and interdependencies, and for negotiations among the formulations that emerge from a spectrum of approaches. "Newly" in quotes, of course, in part because one of Darwin's most famous observations had precisely to do with variety and dense interconnection (his "tangled bank"). That variety has generally been taken to indicate a profusion of species, but it can now encompass a growing number of transgenerational processes as well. The quotation marks are there also because the developments I find most promising today can be traced back at least to the last quarter of the 20th century, and each has ancestors further back. (Readers will also detect suspiciously close ties to the concerns aired in the previous four responses, so in yet another way these problems are not quite "new.")

Here are two such shifts, themselves related: First, theorists are moving away from a narrowly DNA-defined inheritance to more commodious schemes.[8] At the same time more is made of organisms' influence—developmental, physiological, behavioral—on their own evolutionary possibilities.

So one of the major issues facing theorists today is articulating and evaluating the candidates for a heredity no longer confined to the genetic (or even the cellular). In some approaches expansion is minimal, to include Dawkinsian memes, say, or some other channel for culture, while in others, multiple channels are invoked, or heterogeneous systems, constructed niches, a motley of replicators (Aunger, 2000; Jablonka & Lamb, 1995; Sterelny, Smith, & Dickison, 1996). Many of these efforts are documented in the collection, *Cycles of Contingency* (Oyama, Griffiths & Gray, 2001), but the urge itself is considerably broader; certainly it becomes harder and harder to maintain that evolution is best conceived as a matter of changing allele frequencies in gene pools.

At the same time, the habit of seeing populations as shaped from the outside by an independent environment may be weakening as the relationship between organisms and their surrounds is increasingly said to be a matter of coevolution, interpenetration, co-construction, or some other intimately reciprocal relationship, in which organisms are participants in their evolutionary histories (Avital & Jablonka, 2000; Jablonka & Lamb, 2005; Johnston & Gottlieb, 1990; Odling-Smee, Laland, & Feldman, 2003; West-Eberhard, 2003; Wimsatt & Griesemer, 2007).

It is as though once evolutionary theory had elaborated a quite widely-used set of tools, workers realized that these instruments could be deployed beyond their usual ranges, pulling a surprising range of phenomena into evolutionary narratives. In addition, research produced phenomena that didn't fit easily into existing schemes. Whatever the reasons, organismic ontogenies are less apt to be seen as dead-end arrows angling off from classical textbook diagrams' genotype-to-genotype flow through the generations (Griesemer & Wimsatt, 1989). Instead (to phrase it in a way that some would find extreme), ontogenies are the very stuff of evolution, and the exploration of this shift lies largely ahead of us.

From the other direction, from traditional theory's "inside,"

---

[8] For historical background on heredity see Müller-Wille & Rheinberger, 2007.

Thompson (2007) argues that the concerns with developmental dynamics and "internal constraints" on evolution that were aired in the last quarter of the 20th century (see Question 1) have to some extent generalized to a preoccupation with the relationship between self-organization and natural selection. Once more we have a variety of approaches. I suspect a primary question here will be whether we must continue to consider these to be the insides and outsides of evolutionary processes, or whether there is a notion of self organization that can synthesize them.

All of this generates fruitful friction, which in turn generates the next questions, which become important in the crucible of disagreement, challenge, and theoretical consolidation. To some, this may seem an embarrassment of riches, but to me it looks more like bracing vitality.

Apart from the abovementioned sorting out of additional processes of heredity, there will have to be explorations of interactions among them. My own guess is that as this happens the language of channels and information flow will reach the limits of its usefulness (see also Griffiths & Gray, 2001). If, as seems to be the case, monolithic genocentric perspectives on development and evolution are losing their purchase, I wonder whether the infomania of the past few generations will fade too, as more systemic views come to predominate. (They may not; as I noted, these developments include many information-based approaches, though not all of them are what I would dub classical GOFIB, Good Old-Fashioned Infocentric Biology).

Perhaps biology will even rediscover the fleshly creatures (with bodies!) that used to populate our definitions of life. Will a discipline that has become strangely matter-phobic, delineating its subject in terms of algorithms and replicating, substrate-neutral information even as it trumpets its hardheaded materialist credentials (Oyama, 2009a), will this evolutionary biology rediscover a teemingly concrete world full of irritable, motile, reproducing, metabolizing organisms? Will it realize that it was its own unwitting devaluation of matter, of mere phenotypes (mere beings in their surroundings) that necessitated all that information in the first place? For when we turn our gaze to three-dimensional entities at all scales, interacting in space and time, we discover that once we have objects and processes, along with the conditions they require to occur, *we don't need information to represent and make things happen.* Infotalk finally reveals itself to be a denatured, etiolated stand-in for a brimming, squalling world that is more various, and less neatly compartmentalized, than the ones we are used to being shown by our science.

References

Amundson, R. (1994). Two concepts of constraint: Adaptationism and the challenge from developmental biology. *Philosophy of Science, 61*, 556-578.

Aunger, R. (Ed.) (2000). *Darwinizing culture: The status of mimetics as a science.* Oxford: Oxford University Press.

Avital, E. & Jablonka, E. (2000). *Animal traditions: Behavioural inheritance in evolution.* Cambridge: Cambridge University Press.

Godfrey-Smith, P. (2000). Explanatory symmetries, preformation and developmental systems theory. *Philosophy of Science, 67* (Proceedings), S322-S331.

Gould, S. J., & Lewontin, R. C. (1979). The spandrels of San Marco and the Panglossian Paradigm: A critique of the adaptationist programme, *Proceedings of the Royal Society of London B 205*, 581-598.

Gray, R. D. (1987). Beyond labels and binary oppositions: What can be learnt from the nature/nurture dispute? *Rivista di Biologia/Biological Forum, 80*, 192-196.

Gray, R. D. (1989). Oppositions in panbiogeography: Can the conflicts between selection, constraint, ecology, and history be resolved? *New Zealand Journal of Zoology, 16,* 787-806.

Griesemer, J., & Wimsatt, W. (1989). Picturing Weismannism: A case study of conceptual evolution. In M. Ruse (Ed.), *What the philosophy of biology is: Essays for David Hull* (pp. 75-137). Dordrecht, Netherlands: Kluwer Academic.

Griffiths, P. E., & Gray, R. D. (1994). Developmental systems and evolutionary explanation. *Journal of Philosophy, 91*, 277-304.

Griffiths, P. E., & Gray, R. D. (2001). Darwinism and developmental systems. In S. Oyama, P. E. Griffiths, & R. D. Gray, (Eds.), *Cycles of contingency: Developmental systems and evolution* (pp. 195-218). Cambridge, MA: MIT Press/Bradford Books.

Haila, Y. (1997). Discipline or solidarity? Ecology as politics. In P. J. Taylor, S. E. Hafton, & P. N. Edwards (Eds.), *Changing life: Genomes—ecologies—bodies—commodities* (pp. 102-120). Minneapolis, MN: University of Minnesota Press.

Hayles, N. K. (1999). *How we became posthuman: Virtual bodies in cybernetics, literature, and informatics.* Chicago, IL: University of Chicago Press.

Jablonka, E., & Lamb, M. J. (1995). *Epigenetic inheritance and evolution: The Lamarckian dimension.* Oxford: Oxford University Press.

Jablonka, E., & Lamb, M.J. (2005). *Evolution in four dimensions.* Cambridge, MA: MIT Press.

Johnston, T. D., & Gottlieb, G. (1990). Neophenogenesis: A developmental theory of phenotypic evolution. *Journal of Theoretical Biology, 147,* 471-495

Kay, L. E. (2000). *Who wrote the book of life? A history of the genetic code.* Stanford, CA: Stanford University Press.

Keller, E. F. (2000). *The century of the gene.* Cambridge, MA: Harvard University Press.

Keller, E. F. (2001). Beyond the gene but beneath the skin. In S. Oyama, P. E. Griffiths, & R. D. Gray (Eds.), *Cycles of contingency: Developmental systems and evolution* (pp. 299-312). Cambridge, MA: MIT Press.

Lewontin, R. C. (1983). Gene, organism and environment. In: D.S. Bendall (Ed.) *Evolution from molecules to men* (pp. 273-285). Cambridge: Cambridge University Press. (Reprinted with new introduction in Oyama, Griffiths, & Gray, 2001.)

Life & Mind Seminars. Blog on Developmental systems theory and autopoiesis [Internet]-CCNR, University of Sussex, Brighton, UK-[cited 2008 August 27]. Available from:
http://lifeandmind.wordpress.com/2008/08/01/developmental-systems-theory-and-autopoiesis/

Maturana, H. R., & Varela, F. J. (1987). *The tree of knowledge.* Boston, MA and London: New Science Library: Shambhala.

Maynard Smith, J., Burian, R., Kauffman, S., Alberch, P., Campbell, H., Goodwin, B., Lande, R., Raup, D., & Wolpert, L. (1985). Developmental constraints and evolution. *Quarterly Review of Biology, 60,* 265-287.

Moss, L. (2003). *What genes can't do.* Cambridge, MA: MIT Press.

Müller-Wille, S., & Rheinberger, H-J. (Eds.) (2007). *Heredity produced: At the crossroads of biology, politics, and culture, 1500-1780.* Cambridge, MA: MIT Press.

Nelkin, D., & Lindee, M. S., (1995). *The DNA mystique: The gene as a cultural icon.* New York: W. H. Freeman.

Neumann-Held, E. M. (1999). The gene is dead–long live the gene! Conceptualizing genes the constructionist way. In P. Koslowski (Ed.), *Sociobiology and bioeconomics: The theory of evolution in biological and economic theory* (pp. 105-137). Studies in Economic Ethics and Philosophy, Vol. 20. Berlin, Germany: Springer.

Odling-Smee, J., Laland, K. L., & Feldman, M. W. (2003). *Niche construction: The neglected process in evolution.* Princeton: Princeton University Press.

Oyama, S. (1992). Ontogeny and phylogeny: A case of metarecapitulation? In P. [E.] Griffiths (Ed.), *Trees of life: Essays in philosophy of biology* (pp. 211-239). Dordrecht, Netherlands: Kluwer Academic. Reprinted in S. Oyama, (2000). *Evolution's eye: A systems view of the biology-culture divide.* Durham, NC: Duke University Press.

Oyama, S. (2000a). *Evolution's eye: A systems view of the biology-culture divide.* Durham, NC: Duke University Press.

Oyama, S. (2000b). *The ontogeny of information: Developmental systems and evolution* (2nd ed., revised and expanded). Durham, NC: Duke University Press (originally published 1985, Cambridge University Press).

Oyama, S. (2002). The nurturing of natures. In A. Grunwald, M. Gutmann, & E. M. Neumann-Held (Eds.), *On human nature: Anthropological, biological, and philosophical foundations* (pp. 163-170). Wissenschaftsethik und Technikfolgenbeurteilung Band 15. Berlin: Springer Verlag.

Oyama, S. (2006a). Boundaries and (constructive) interaction. In C. Rehmann-Sutter & E. M. Neumann-Held (Eds.), *Genes in development. Re-reading the molecular paradigm* (pp. 272-289). Durham, NC: Duke University Press.

Oyama, S. (2006b). Speaking of nature. In Y. Haila & C. Dyke (Eds.), *How does nature speak? The dynamics of the human ecological condition* (pp. 49-65). Durham, NC: Duke University Press.

Oyama, S. (2009a). Compromising positions: The minding of matter. In A. Barberousse, M. Morange, & T. Pradeu (Eds.), *Mapping the future of biology: Evolving concepts and theories* (pp. 27-45) Boston Studies in Philosophy and History of Science. Berlin: Springer Verlag.

Oyama, S. (2009b) Friends, neighbors, and boundaries. *Ecological Psychology,* special edition, *21,* 147-154.

Oyama, S., Griffiths, P. E., & Gray, R. D. (Eds.). (2001). *Cycles of contingency: Developmental systems and evolution.* Cambridge, MA: MIT Press/Bradford Books.

Poirier, P., Faucher, L., & Lachapelle, J. (2005). The concept of innateness and the destiny of evolutionary psychology. *Les Cahiers du LANCI, 4,* 2005-01, 1-34.

Pradeu, T. (submitted). The organism in DST.

Sterelny, K., Smith, K. C., & Dickison, M. (1996). The extended replicator. *Biology and Philosophy, 11,* 377-403.

Thompson, E. (2007). *Mind in life: Biology, phenomenology, and the sciences of mind.* Cambridge, MA: Belknap/Harvard University Press.

Tooby, J. & Cosmides, L. (1992). The psychological foundations of culture. In J. H. Barkow, L. Cosmides, & J. Tooby (Eds.), *The adapted mind* (pp. 19-136). Oxford: Oxford University Press.

Varela, F. J. (1979). *Principles of biological autonomy.* New York: Elsevier North Holland.

Weber, B. & Depew, B. (Eds.) (2003). *Evolution and learning: The Baldwin Effect reconsidered.* Cambridge, MA: MIT Press.

West-Eberhard, M. J. (2003). *Developmental plasticity and evolution.* Oxford and New York: Oxford University Press.

Wimsatt, W. C. & Griesemer, J. R. (2007). Reproducing entrenchments to scaffold culture: The central role of development in cultural evolution. In R. Sansom & R. Brandon (Eds.), *Integrating evolution and development: From theory to practice* (pp. 227-323). Cambridge, MA: MIT Press.

17
David C. Queller

Harry C. and Olga K. Wiess Professor in Natural Sciences
Department of Ecology and Evollutionary Biology, Rice University
USA

---

1. Why were you initially drawn to discussions and research on evolution (or evolutionary aspects of your field)?

My father was a medieval historian, and I suppose that gave me an early exposure both to historical contingency and to social and moral issues. For example, the medieval Venetians had long been reviled as unusually selfish villains for their role in diverting the Fourth Crusade to sacking the Christian city of Constantinople. My father argued that the Venetians were basically normal people, complex mixtures of good and bad, and that their behaviour in this event followed from a series of contingencies that put them in a difficult position. I toyed with the idea of doing history myself, and indeed majored in the history and philosophy of science at the University of Illinois. But, really it was just a way to try to balance my interests in things social and scientific and to have maximum opportunity to take courses in both.

Ultimately I decided that history was just a little too contingent. Paul Valéry wrote that "History is the science of what never happens twice", which is to say, not really science at all. I preferred, in the end, things that could happen at least twice, and could therefore be tested. This was taking things a bit backwards. Scientists often develop an interest in the history of their field as they age, but I read the history of evolutionary biology before I became an evolutionary biologist.

What were the attractions of evolution? First, it was a big theory. Darwin was the Newton of the living world; he explained how living things came to be the way they are. Adaptation by natural selection is what animates living matter, what makes it different from mere chemistry and physics. Moreover, you didn't need to be a math geek to understand it (though the theory is much more subtle and complex than most appreciate). And the sheer hubris of Darwin's enterprise was appealing. He started from the patently obvious, that selection should work if there is heritable variation that affects survival, and extrapolated to the outlandish, that we could explain the traits of monkeys, mackerels, maples, and mushrooms – and that we were cousins to them all. Here was history, with its interesting contingencies, but also with a simple law driving it.

When I enrolled in 1976 in the PhD program at the University of Michigan, my interests began to turn to social evolution. This was the year Richard Dawkins' *Selfish Gene* was published, and just the year after E.O. Wilson's *Sociobiology*. My choice of Michigan was truly fortunate (my alternative was studying at Harvard, maybe with Stephen Jay Gould – Harvard had rejected my application, but reconsidered when I obtained an NSF fellowship that would pay my way). Michigan had Richard Alexander, who was one of the pioneering thinkers in sociobiology. In my first semester I was immersed in his course, knowing the impossibility of his demand that we understand every sentence of the literature readings he assigned, and nevertheless trying to do so. At the same time, I was taking a seminar from John Maynard Smith, who was visiting for the fall semester, getting an introduction to game theory and the evolution of sex. In the following fall semesters, we had similar visits from Bill Hamilton (who later returned to a permanent position) and George Williams, so I had the unusual opportunity of learning directly from not just one, but from several leaders in the field.

I think social evolution does have a special appeal. We humans are extremely social animals, so exploring the biology of conflict and cooperation, of sex and parent-offspring relations, cannot fail to be interesting. To some degree, we see ourselves in our study animals and we see our study animals in ourselves, though we need to be careful to keep those separated in our science. What could be more interesting than understanding how social behaviours evolved and perhaps how moral systems came to be? Perhaps even more important, because social evolution had not been thought about very rigorously before the 1960's, there were countless unsolved problems to work on.

2. What does your work reveal about biological evolution that other academics, citizens, philosophers or biologists typically fail to appreciate?

Speaking not just for my own work, but also for others in the field of social evolution, I think we have discovered a wonderfully complex landscape of selfishness and cooperation in the natural world. Depending on their background, some people fail to appreciate the importance of selfishness and others fail to appreciate the importance of cooperation.

The main contribution of the revolution in social evolution in the 1960's and 1970's, led largely by people I have mentioned above, is to cement the conclusion that all adaptations, including social adaptations, rest on a form of genetic selfishness. A trait can persist only if genetic alleles contributing to that trait leave more copies of themselves than alternative alleles making different traits. This seems like simple uncontroversial Darwinism, and it is supported by simple mathematical population genetics, but thinking about social adaptations had sometimes been rather different, often assuming that selection would produce individuals that did things for the good of the group or for the good of the species. Therefore, much of social behavior had to be rethought and reinterpreted in terms of what was good for individuals. For example, though males and females sometimes cooperate in rearing young, there are potential conflicts. Each might do better if it could induce the other raise the young alone, and males might be particularly prone to skip out on care if they had opportunities to attract other females. Or a female might do better if she copulated with a better male than her social mate, though her mate would be selected to try to prevent this. Everything had to do with what was best for individuals (and their genes) and nothing with what was best for the species.

Here is one example of this mode of thinking from my own thesis work. Flowers are often showy. This serves to attract pollinators that are necessary to move pollen to a stigma, where it can germinate and fertilize an ovule. Many plants, and even their individual flowers, are hermaphroditic, having both male and female parts, but it is still often necessary to move the pollen from one plant to another. I showed that the floral displays of a milkweed were not selected to maximize seed production, which is what would help this species in competition with others. Indeed, most seeds are aborted because the plant cannot afford to raise them. Much smaller displays would suffice to fertilize all the ovules that the plant could afford to develop into seeds, so why have extravagant displays? The showy displays served instead to increase the genetic success of these hermaphroditic plants through the transport of their pollen to other plants, enabling them to outcompete other plants. Instead of making as many seeds as possible, which would be best for the population, these plants spend their resources trying to outcompete each other though their male functions.

On the other hand, the good-of-group viewpoint was not entirely wrong. Social insect workers spend their lives helping the queen to reproduce, forgoing their own reproduction. So the problem became how to explain these truly altruistic behaviours in a model of genetic selfishness. The answer of course was kin selection, the subject on which I have spent the better part of my career, showing that it operates not just in social insects, but in organisms as different as plants and amoebas. The idea, due to Bill Hamilton, is that an allele can reproduce not just through helping to produce offspring, but also through helping other relatives, because other relatives will share the allele with a specifiable probability, for example 1/2 for full siblings. A social insect worker that reared enough siblings could compensate for not raising any of its own offspring.

This selfish-gene viewpoint is somewhat old news now, but it was surprising and somewhat shocking at the time, and still surprising and shocking for those who encounter it for the first time. One common reaction to this news is to think that the world is irredeemably selfish. Against this background, the point that needs to be stressed now is how common and important cooperation is. The altruistic social insects are far from the only examples, a point I will return to. But perhaps more significantly, the social insects point us to the depth of cooperation, in the following way. It has long been argued that at least the most elaborately social of the social insects, taxa like honeybees or leaf-cutter ants, have colonies that are essentially organismal in structure. We sometimes call them superorganisms, or organisms that are above the level of normal organisms. The parts of the colony interact are adapted to work in intricate, cooperative ways for the good of the colony, much as the parts of an organism's body do. They must have evolved these nearly completely cooperative organismal structures out of the kinds of societies we see in simpler social insects, like paper wasps or sweat bees, which show considerable conflict between individuals.

If organisms evolved out of groups of individuals in the case of social insects, what about conventional organisms? They turn out to involve similar histories. Multicellular organisms are groups of cells that have evolved to cooperate. In most cases, the cooperation is easier to evolve because the cells are genetically identical, provided they have evolved ways to keep out intruders. But it was still a process of evolution by kin selection. At the next level down, the eukaryotic cell is also historically an assemblage of multiple individuals, with mitochondria being derived from bacteria and chloroplasts for blue-green algae. In this case, the organism did not evolve via effects on kin, but through mutual advantage to the associating species. A similar process must have been involved in the origin of the first cells, where different kinds of components were packaged together and had to evolve to cooperate. In other words, the conclusion that high degrees of cooperation are scarce in the natural world is an error. Each organism is an evolved cooperative entity of incredible intricacy, but one that is so deeply cooperative and integrated that we tend to ignore the fact that it even has subunits.

The social insects also show the result of cooperation. Although there are relatively few highly social insects, they have been very successful. Their cooperation allows them achieve things that other insects can't and to dominate their ecosystems. Likewise, the other transitions to new levels of organisms are relatively rare, but have been profoundly important. What could be more important in the history of life than cells, eukaryotic cells, and multicellularity? Each new level of cooperation is hard to achieve, but once achieved, can have extraordinary potential for success.

However, there are also more obscure transitions to new levels of organisms. Recently, I argued, together with my collaborator (and wife) Joan Strassmann, that we ought to define organisms along these lines – they are compound entities that have near universal cooperation with very little conflict. We argued further that, if we use this definition, there are some entities that are not normally considered organisms that probably ought to be placed in that category. One example that many accept is social insect colonies, at least among advanced species with little conflict. Other kinds of unconventional organisms might include certain mated pairs, such as in anglerfish where a tiny male fuses for life with a much larger female. Certain microbial groups can be viewed as organismal, including social amoebae (more on these below) and perhaps some infective organisms. We know from the eukaryotic cell that mutualisms can lead to organisms, and it seems likely that some other mutualisms can also be viewed as organismal. These unconventional organisms have not had the same huge effects on life as the major transitions discussed above, but they show that the major transitions are just part of the cooperative landscape.

So, my main point here is that cooperation is both more pervasive and more important than we might have thought in the early days of the selfish gene. It is remarkable how often selfish genes can lead to cooperation. But let me close this topic by turning once again to the selfish part. My identifications of unconventional organisms in the previous paragraph, for example between male and female anglerfish, are tentative; they would be overturned if further study shows sufficient conflict between the components. But we cannot demand a complete absence of conflict, for even conventional organisms have some conflict among their parts. Many of the examples come from the murky realm of molecular evolution, where transposable elements or restriction modification systems act to propagate their own sequences, sometimes at some cost to the organism that bears them.

But my favourite example, genomic imprinting, is more tangible. In a diploid organism, each genetic locus has an allele inherited from the mother and another from the father, which I call matrigenes and patrigenes. Usually these alleles work harmoniously together, but David Haig has shown that sometimes they act in opposition, in ways expected from kin selection theory. When an offspring obtains resources from its mother, selection weighs not only the benefits it gets, but also the costs it imposes on the mother's other offspring, its siblings. Suppose mothers sometimes mate multiply so that siblings are often half siblings, related through matrigenes but not through patrigenes. Patrigenes then have no genetic stake in these half siblings, but matrigenes do, so one would predict that patrigenes would be less averse to imposing costs on siblings. Several facts agree this scenario. First, imprinted genes (those that act differently in matrigenes and patrigenes) are most common in mammals and seed plants, where offspring can affect the amount of resources obtained from the parent. Second, those genes are commonly expressed in embryos or in placentas and endosperms that are genetically similar or identical to the embryo and function as food gathering tissues. Third, when the effect is known, the patrigene acts more to increase food acquisition than the matrigene. Such wars among genes of the same organism remind us that, even within organisms where cooperation is most extreme, the pull of selfishness remains and some genes evolve ways to go their own way.

3. What, if any, practical and/or social-political and/or moral obligations follow from your work on evolution?

My impulsive answer to this question is "none", at least with respect to the social-political and moral angles. Assuming we can successfully understand the evolution of animal and human sociology, does that tell us how we should behave? Most philosophers agree (and how often can that be said about anything?) that "is" does not imply "ought". It must be even more difficult to get to "ought" from an evolutionary "was". It is evolutionarily natural in some sense for humans to run about naked, to live in small bands, and to hunt and gather, but that does not make those activities morally good. Conversely, it is evolutionarily unnatural to fly in planes, to live in cities, to cultivate crops, and to read, but that does not make these activities bad.

Yet, philosophers notwithstanding, the view that there is some kind of leap from "is" to "ought" is common. Sociobiologists who have tried to make that leap have been justly criticized for it. Or at least they have been criticized when that leap leads to distasteful "oughts". To be consistent, we should also criticize those who try to align "is " and "ought" in more palatable ways. For example, though they might not put it quite this way, it is hard not to see David Sloan Wilson's crusade for group selection as partly motivated by a desire to justify good behaviour, and Joan Roughgarden's revisiting of sexual selection as partly motivated by a desire to justify variant sexual roles. Most sociobiologists steer clear of such thinking most of the time, but we still get criticized because our "is" might make someone else might leap to "ought". So I guess our main moral obligation is to let people know that no moral obligation follows from our findings.

A more limited goal is for sociobiology theory to tell us what human nature is. I am sceptical even about this more restricted goal. We do not need a biological theory to tell us what human nature is, for example whether humans are good or evil. Human nature is what it is; it is an empirical question not a theoretical one. Human nature encompasses Hitler and Gandhi alike, with most people, like my father's Venetians, falling somewhere well between those poles. A theory cannot tell us which pole is dominant.

What theories can do, if we are lucky, is explain what we see. At best, such understanding might help us shift a bit (towards

Gandhi, I hope). I am getting a bit out of my specialized expertise here; a life spent studying plants, insects, and amoebas hardly qualifies me as an expert, except perhaps in knowing what all these other taxa lack. But the work of sociobiologists, economists, and anthropologists does seem to show that there is something different about human cooperation. Obviously we have high intelligence, but it is not that we have simply figured out that it is good to cooperate. Instead it appears that our intelligence, and especially our language abilities, have endowed us both with a complex material culture, which gives us something to cooperate about, and also the ability to negotiate the complexities of making cooperation pay.

With respect to the first point, the increased complexity of human culture can lead to specialized abilities in particular learned tasks. This can turn reciprocity from a weak trading of a common service – you pick my nits and I'll pick yours – into a powerful one where each partner exchanges its special skill for others that it is not so good at. Obviously this is extremely important in modern monetary societies, where I can, through the medium of money, trade my evolutionary biology skills a chance to enjoy Yao Ming's extraordinary basketball talent. But it seems likely to have also been relevant in hunter-gatherers. Some individuals will be better making weapons, others at hunting deer, others at fishing, others at gathering tubers, others at making fire or building shelters or at preparing medicines. The ability to exchange different services is what makes between-species mutualistic interactions powerful, and humans are unique in the way they achieve it within a species.

So specialized knowledge and skills make exchange more valuable. Enhanced intelligence plays a second role in allowing better tracking of the behaviour of partners. Differential treatment of cooperators and cheaters requires that they be recognized and remembered, either from your own interactions with them (leading to direct reciprocity) or through knowledge of their interactions with others (leading to indirect reciprocity). Human language puts this into a whole new realm because people learn about the behavior of potential partners, their reputations, not just by observing them but by hearing about them from third parties.

Finally, humans may also cooperate because of group selection, especially on culture. Perhaps because of the importance of imitation and learning, human societies develop cultural norms for how things are done, and this includes social norms. Anthropologists have pointed out that since such norms homogenize traits within groups, they weaken the power of selection within groups relative to between groups. Groups with cooperative norms can replace those with uncooperative norms without too much danger of opposite within-group selection; even migrants from non-cooperative groups can be assimilated into the cooperative norms. To come full circle with a little speculation, the importance of norms in human evolution might even account for our desire to equate "is" with "ought". In most human cultures, the norm is in fact prescriptive for how we ought to behave.

4. What do you see as the most interesting criticism against your position in the biological or philosophical discussion of evolution?

Stephen Jay Gould has laid out the broadest challenges to my general field, some of them more interesting than others. His political views on sociobiology are not the most compelling of these. He argued that sociobiologists infuse various biases – capitalist, sexist, racist – into their work, building their biology as a distorted image of their own society. I'm not claiming that this is never true, but the assertion seems way overblown. Indeed, leaving aside the capitalist aspect, I suspect there are few human societies less disposed to these biases than the modern liberal democracies from which sociobiology emerged. Are there any other societies where so many at least take it as an ideal that men and women should be treated equally and allowed to chose their own roles? Are there others where discussion of racial differences approaches a taboo? And how to explain the importance in sociobiology of kin selection, when it came from modern societies where kinship ties may be less important than in almost any other time and place?

Again, I am not arguing that we are free of bias. What is most important is that we, like other scientists, use mechanisms to try to purge bias and approach the truth. In science, you get points for knocking the other guy down and you get points for coming up with better ideas. This is exactly the point that creationists fail to understand when they complain that Darwinism is supported only by a cabal of scientists bent on promoting their godless, humanistic biases. Some of us are godless and even more of us are humanistic, but nearly all of us are ambitious, and there could be no greater prize than to prove Darwin wrong. Likewise, though the stakes are smaller, we get rewards for knocking down the weak sociobiological edifices, and for getting the evidence right and building better structures in their place. Unless there is an overwhelming politically oriented cabal out there, inappropriate biases will tend to be weeded out.

For a more specific example, consider my old field of social insect research. Social insects are highly altruistic, and therefore provided a challenge to an individual-centered selection framework. However much sociobiologists might have been biased in favor of selfishness, no one took the approach of saying that no, the social insects never behave altruistically. Instead, Hamilton rose to the challenge of explaining such altruism via inclusive fitness. This was not a mere band-aid applied to a hemorrhaging theory in a desperate attempt to save it. It did patch that particular wound, but not in an *ad hoc* way. The idea was shown to be mathematically sound, and indeed to be a logical conclusion of selection theory. And more important, as expected of a good scientific theory, it made additional predictions. For example, its application to the strange haplodiploid genetic system of ants, bees, and wasps led to some rather odd predictions that workers should favor more female-biased brood ratios than queens. Those predictions have been very successful. In wielding inclusive fitness theory, we have come up with a view of social insects that is in pretty stark contrast with a capitalist or conservative political ideology. We see social insects as highly cooperative and usually working for the good of their group, with only minor conflicts. We view the males of ants, bees, and wasps, as rather uninteresting, because it is only the females that are cooperative. And most of us now believe that, at least in the more advanced social insects, the workers are basically in charge. They determine, by feeding, which individuals become queens versus workers, and they determine, by egg eating or aggression, whether other workers will be allowed any reproduction. So somehow, in spite of whatever insidious biases we may have, we have arrived at a result that is pro-cooperation, pro-female, and pro-worker. Scientific ideas take on lives of their own, beyond whatever ideology might have lurked in the back of the minds of their originators.

It is tempting to think that Gould's opposition to sociobiology was completely political in nature, and therefore easily dismissed. It is true that he did not like some of what he saw as the political baggage of human sociobiology, and that distaste may have flowed over into sociobiology as a whole, and then even into adaptationism in general. I have enjoyed taking some pokes at that viewpoint, and especially at those who were too readily swayed by Gould's formidable writing skills (see The Spaniels of St. Marx, Quarterly

Review of Biology 70: 485-489). But there are serious issues here, ones that start from a broad view of adaptation rather than from a parochial view of sociobiology.

Darwin described evolution as "descent with modification". These two nouns summarize Darwin's two main achievements. First, he built a compelling case for common descent. Common descent was not an especially new idea, but Darwin gathered a huge amount of evidence, particularly from homologies and biogeography, that supported a branching model of descent. Second, he showed how, in spite of this common descent, species could become different, via his mechanism of natural selection. Both elements remain essential in any modern form of Darwinism. We need both history and design, both chance and necessity. But different researchers give them different emphasis.

Sociobiology grew out of the camp that is more attuned to modification, design, and necessity. Its main question concerned what kinds of adaptations we expect out of natural selection when individuals affect others or, to put it in group-selection terms, what kinds of adaptations we expect when the good for the individual and the good for the group do not coincide. Following the decisive shift to an individual inclusive-fitness-centered approach, the whole range of social adaptations came up for reinterpretation, and the job was to see if adaptations that had been attributed to group or species advantages could be understood instead in terms of individual advantages. In the contest between levels of adaptation, issues of constraint and non-adaptation were often set aside. Constraint was not rejected, but it was often ignored to see how far one could get with assuming things could change freely.

Gould, more than most evolutionary biologists of the time, was more attuned to the descent side of Darwin's descent with modification, to history and to the role of chance. We detect common descent through traits that have not changed much, and these seemed central to Gould. Much of his thought seems to converge on the ideas that constraints are relatively strong, and that even when change does occur, it is unpredictable and contingent. Indeed, one of the reasons to be sceptical about Gould's views on political influences is that, in spite of Gould's leftward political stance, his evolutionary biology seems fundamentally conservative. When he argues that change is difficult, that changes often entail deleterious side-effects, and that change builds idiosyncratically on whatever idiosyncratic system that is already there, he sounds like the biological incarnation of Edmund Burke, the patriarch of English conservatism, who made similar arguments about society. An exception is that large changes were anathema to Burke but have some appeal to Gould, but nevertheless a conservative streak is unmistakable in Gould's biology. There's nothing wrong with that because Gould's biology does not need to be chained to his politics, but then he should have appreciated that he is not the only one capable of slipping such shackles.

One reason that I think that Gould's anti-sociobiology stance was not entirely political is that his views on such issues preceded, at least in part, the sociobiology debates. The theory of punctuated equilibrium, published in 1971, took a strong line on stasis and constraint. Gould believed that most of the time, species did not change much, largely because traits were too inextricably tied to one another, through pleiotropy, linkage, and epistasis. Changing one trait in a good direction would often be blocked by the deleterious side-effects on other traits. He thought that serious change needed a genetic revolution in peripheral speciating populations, with drift in these small populations shaking them out of their tightly coadapted states. What happens to adaptation under this view? Within populations, adaptation happens in very short periods, and is ineffective much of the rest of the time. As a consequence, much of what accounts for the traits we see in nature may be a process or sorting among more-or-less fixed species. He explored the issues of adaptation and constraint further in *Ontogeny and Phylogeny*, which though published shortly after the initial sociobiology brouhaha, must have been largely a pre-sociobiology work. Again, he pursues the line that evolutionary change is often difficult, though there may be some comparatively easy changes, such as through simple allometric changes or shifts in developmental timing such as neoteny.

Besides arguing that change is difficult, Gould argued that it was contingent and unpredictable. Some changes depend on the lucky mutation that did not have to have happened, or on the lucky gene combination engendered by genetic drift. Others are contingent on a lucky escape through a mass-extinction event that no one was adapted for, and on the bad luck of those that, by not making it through, vacated ecological niches. And all depend idiosyncratically on the pre-existing architecture that is unique to each organism. Gould's evolution approaches the history of Valéry, where nothing happens twice.

If evolution is hemmed in on all sides by constraints, and if changes are highly subject to the vagaries of chance events, then adaptation would be much weaker than we often assume it to be. I think the evidence is shows that adaptation is actually quite pervasive, but there is no space here to rehearse all the arguments for and against Gould's views. Suffice it to say that, though some aspects may be wrong and others pushed too far, they are interesting, and to the extent they are right, they make a difference for how we view evolution.

5. With respect to present and future inquiry, how can the most important problems concerning evolutionary theory (or evolutionary aspects of your field) be identified and explored?

I am not one for forecasting the future. Almost anyone can pick some of the important problems that we need solved. The trick is picking aspects of these problems that actually can be solved. Some problems fall to new technology, and those are sometimes predictable; it is clear that new mass-sequencing technologies are going to have a great impact in various ways. Others succumb through the importation of ideas from other fields, and that's harder to predict. I hope that systems biology and network approaches will be able to help with the issue of how evolution works on traits that are tied together. But other major advances simply require some novel and clever insights, and if I knew what those insights would be, I would be publishing them in the present rather than forecasting them for the future.

One thing can perhaps be said about these new insights. If you really want to know what the future will be, ask a young scientist. Ask someone who is fresh out of grad school and has heard the best that we have to offer, but whose neural connections are still loose and flexible. Of course, one cannot know which one to ask – most will become normal scientists and follow the herd – but a few will make the leaps that define the next era. They are probably already doing it and I am too hidebound, or too busy with my own concerns, to see it.

I can speak with confidence only about the paths I am following. One of the main ones, again with Joan Strassmann, is to look at real selfish genes. The odd thing about the paradigm of the selfish gene is that it has been largely carried out in the absence of any knowledge of the underlying genes. We have studied traits and counted gene copies but we have not worried much about the genes causing those traits. In order to be able to study sociality at the genetic level, we decided to switch away from social insects, which are difficult to study genetically (although progress is being made with honey bees). We opted instead to switch to a model organism that had not been widely recognized as a good social organism, the social amoeba, *Dictyostelium discoideum*. Like other species in its ancient genus, the amoebas are single cells that are normally solitary. However, when their bacterial food supply runs out, they switch into social mode. They gather together in groups of thousands, and build multicellular fruiting bodies. In these fruiting bodies, about 20% of the cells die in the process of forming a stalk, which facilitates the dispersal of the rest of the cells, which differentiate as spores.

D. discoideum is interesting at two levels. One is that it offers a social system that is very different from those that stimulated the theory of social evolution, so it supplies independent tests of those ideas, such as the importance of genetic relatedness. But we can also exploit the previous mechanistic work on this species: a sequenced genome, extensive knowledge of genes, proteins, and pathways, and the numerous techniques that have been developed. Taking such a leap is best done with collaborators, and we are fortunate to have found very good ones. We have been able to show that cheater mutants are very common, and that they can select for resister mutants. We have shown how high relatedness selects against bad cheater mutants. We are finding kin recognition genes, including a green-beard locus. One of our most interesting findings is that pleiotropy, one of Gould's constraints on adaptation, can actually help with the evolution of cooperation, because it is a constraint on selfish cheater genes. Finally we are curious whether sequence data will show that social genes evolve rapidly in conflict-driven arms races.

Our work in this vein is, I think, part of a larger trend in sociobiology. In the early days it was sufficient to come up with a novel ultimate explanation of some behaviour that was already known. Now much more work is being done on proximate mechanisms. Some, like us, exploit the ease of manipulation of microbial systems. Some are studying hormones or the chemical nature of signals. Others focus on reproductive physiology or gene expression patterns. Still others study genetic effects at a quantitative genetic level or take a phylogenetic approach. In short, we are fleshing out the adaptive accounts with the details of structure, function and history. And that is the way to find out how much the proximate details matter.

Acknowledgments: I thank Joan Strassmann and Gry Oftedal

for comments on the manuscript.

18
Michael Ruse

Lucyle T. Werkmeister Professor of Philosophy and Director of the Program in the History and Philosophy of Science

Florida State University, USA

---

1. Why were you initially drawn to discussions and research on evolution (or evolutionary aspects of your field)?

I am a historian and philosopher of science, with my main focus being on evolutionary theory. I am not a scientist and, as it happens, although my undergraduate degree is in mathematics, and at high school in England I did a lot of physics and some chemistry, I have never taken a biology course in my life. So we need to start there, with why it was that I happened to come to study and write on the fundamental idea in the biological sciences.

The answer I am afraid is really not very exciting or noble. I was born in England in 1940; I went to the University of Bristol in 1959; and emigrated to North America (Canada) in 1962. I had taken some philosophy courses as an undergraduate and on the very first day of class I knew already that that was where I wanted to spend the rest of my life. There really were people who worried about whether they are awake or asleep and if the world exists behind them when they are not looking and if it is possible that everything is doubling in size every second!

Intending to be a professor, I went to graduate school until 1965 when, sick of courses and the courses (and their teachers) sick of me, I dropped out and got a temporary job teaching philosophy at the University of Guelph in southern Ontario. Having to get up in the morning and having to take on some responsibility for what I was doing turned me around (as my parents and headmaster had always told me) and the prospect of a permanent job at Guelph loomed. But to get this, I needed a doctorate. Bristol was prepared to take me in and thus it was that around 1966 I was looking for a PhD thesis topic.

I knew that I wanted to work in the philosophy of science – writing on the history of philosophy never attracted me (today, I could certainly imagine wanting to work on Aristotle), and more conventional or central areas of mid-twentieth-century analytic philosophy had no big magnetic draw. Today I might be tempted by some area in moral philosophy, but back then it was all very technical with no real interest in actual moral issues. The Vietnam war was going on, blacks were rioting in the streets, everyone – except me – was having sex and we would sit around the seminar room taking as our moral paradigm: "You ought not write on the pages of library books." The same irrelevance was true of the philosophy of mind, dominated as it was by the too-clever vacuity of Gilbert Ryle and his Oxford groupies. Today, thanks to cognitive science, it seems to me to be very exciting – although, I am not sure that I would want to work in an area where you cannot even start to solve the main question, namely what is mind.

So philosophy of science, but what aspect of that? For my master's thesis I had written on paradoxes of confirmation and that was enough of that sort of thing. Physics was very technical and in any case I wanted to move away from my past. I was looking for an area, not very well mined, with some literature, preferably not very good. My intended supervisor suggested that evolutionary theory might be a good candidate. I knew nothing about the subject. I guess I was an evolutionist – although I came from a Christian background, it was Quakerism, which is not a very biblically oriented version of that religion, so going with science over Genesis was no problem. It was just that I knew nothing about the field – I think in fact that, had I been tackled, I would have been a Lamarckian believing in the inheritance of acquired characteristics. But ignorance and indifference was the main mark of my thinking rather than deliberate embracing of what I now know to be false.

I went to the most obvious source for a middle-class Englishman of my generation, namely the Penguin paperback on the theory of evolution – in previous years I had gone to Penguin for information on ballet and opera and that sort of thing, so this was a natural thing to do. The evolution Penguin was written by John Maynard Smith – and, like my encounter with philosophy some years previously, it was in love at first word. I thought that evolution was just the most exciting idea I had ever encountered. I want to make clear that I never then, or since, wanted to be a professional scientist working on evolution. I had had enough of mathematics so I did not want to do theoretical work. I truly do not care enough for the outdoors to be an empirical biologist. When my friends insist on taking me to their field stations, I find that one turtle a summer is enough for me and then it is back to camp for a good detective story and a bottle of beer.

I want to make it clear also that I was not looking for a new religion, a kind of scientific substitute for the Christianity of my childhood. I think that this is what evolution has always been for my good friend Edward O. Wilson (and many others through the years) but it was not for me. Rather it was that evolutionary theory seemed such a great idea and full of interesting philosophical questions – about the nature of laws, about explanation and prediction, about testing, and on and on. John Maynard Smith – whom I got to know a bit over the years – opened up that world for me, and I am eternally grateful.

At the end of his book, Maynard Smith recommended some of the classics – Theodosius Dobzhansky's *Genetics and the Origin of Species*, Ernst Mayr's *Systematics and the Origin of Species*, George Gaylord Simpson's *Major Features of Evolution*, and others. I read them all! I also read the philosophical literature on the topic of evolution and the more I read the more it confirmed my belief that it was not very good and that I could do better. It was then that I discovered that I was not alone in this belief and that there were a (small) number of junior scholars, mainly in America – David Hull stands out preeminently – who shared my enthusiasm for evolutionary ideas and my prejudices about the quality of the then-existing work.

And the rest, as they say, is history. I started writing – the first draft of my thesis was done in six months and, because it would be another two years before the regulations would allow me to submit, I started submitting articles for publication – by 1970, I had ten publications and each was a chapter of the thesis. Then it was all rewritten as a short text, *The Philosophy of Biology* (in truth, *The Philosophy of Evolutionary Biology*), my first book published in 1973. I should say that in those days, although I wrote quickly when once I was started, I used to find it a very stressful experience putting pen to paper. (Actually pencil – I did not learn to type until I went back to high school in my late forties.) It was not until around 1980 that I discovered that writing held no terrors for me and that I would not lie awake the night frightened before I was to start on something significant.

The rest was history in another sense also. The dominant philosophical work by the late 1960s was Thomas Kuhn's *The Structure of Scientific Revolutions*. For all my education – six years in total in graduate school – I am really a bit of an autodidact. Teachers were important for me but never in the sense that I felt that they were wise people at whose feet I must sit and absorb the truth. My learning has come from books, and none was more important than Kuhn's. I had come from the formal background of logical empiricism – chief gurus Carl Hempel and Ernest Nagel – using logic and the like to analyze science, something done in a very prescriptive fashion (that is having a preset ideal and then seeing if the science measures up). That was the philosophy of my little first book. I did not agree with Kuhn's philosophy – at least I did not agree with it then, although through the years I have absorbed more and more of the ideas – it seemed to me to be too descriptive and too subjective and anti-rational. But I saw it as a powerful vision that needed to be understood if it was to be challenged.

More than this, I learned from Kuhn that if you are to do the philosophy of science properly, then you had better know the science. And if you want to work on issues like theory change, you had better know the history of science properly. Not just from the quick reading of one or two secondary sources, but by working hard as a historian. I had by then read Darwin's *Origin* – I think it was one of Maynard Smith's recommendations – and I have always loved the Victorian era – Captain Marryat's *Children of the New Forest* was a childhood favorite, and then Dickens, Wilkie Collins, Trollope, and the like — and so I needed no further encouragement. (Not George Eliot. I did not care for her work then and I do not now. *Middlemarch* strikes me as one of the most depressing novels I have ever read. Minor Dickens, *Dombey and Son*, still has me in its thrall.)

In fact, the first paper I ever had accepted for publication was on the Darwinian revolution and why it did not fit Kuhn's theory of theory (or as he calls them, paradigm) change. Proudly, I sent it to the leading historian of evolutionary biology John Greene, and his shriek of agony over a job done so badly could be heard from Connecticut – his home state – to Ontario – my home province. I realized that I must do better and to this end I spent my first sabbatical, from 1972-1973 at the University of Cambridge, where I listened in on people like Robert M. Young and Martin Rudwick (not to mention mixing with the then-graduate student Roy Porter) and spent much time in the Darwin Archives in the University Library.

The end result was my second book, appearing in 1979, *The Darwinian Revolution: Science Red in Tooth and Claw*. I had intended to have a whole chapter showing why the revolution did not fit with Kuhn but at the last moment I dropped it. No one has ever said: "Great book, but why don't you have a chapter on Kuhn?" I wrote it as a kind of overview, the sort of book I could have used ten years before. I also wrote it as homage to Kuhn, who had before *Structure* written *The Copernican Revolution*. I never sent a copy to Kuhn – I did not want to seem like a suck – and have long regretted not doing so.

A good story about the subtitle is that it was supplied by one of the supposedly anonymous readers for the publisher, the University of Chicago Press, in real life David Hull. Several years later, I was in a conversation with Ernst Mayr and the title of my book came up. Mayr was rather given to judgments. "It is not a bad book even though it does not have enough about German thought" – Mayr, I should say was German-born. "However, your subtitle is really pretty silly. I was talking to David Hull and he agrees with me." The whited sepulcher. I have been waiting thirty years to get back at David over that.

So there I was around 1980, historian of evolutionary biology. I thought that I was no longer a philosopher but I guess the golden thread was never broken and increasingly in the subsequent years I found myself being sucked back into philosophy. But I think in many respects I am and have long been a historian of ideas of the Arthur Lovejoy or Isaiah Berlin kind. Wanting to turn to history but in order to solve philosophical problems. That is what I like to do and in respects what I do best.

2. What does your work reveal about biological evolution (or evolutionary aspects of your field) that other academics, citizens, philosophers or biologists typically fail to appreciate?

I see what I have done in my lifetime in the history and philosophy of science as falling into two categories. First, there has been what one might call the social. There were huge numbers of jobs for philosophers in the 1960s – this was the time of the Beatles and flower power and Eastern mysticism and that sort of thing. Every kid at university wanted to do philosophy. Then in the 1970s, dreams of jobs started to fade and no one wanted to do philosophy. It was all medicine or economics – later, computers – and that sort of thing. Although my department had a PhD program, there were no jobs for the students. I found this really depressing and in the end – although I mentored a couple of people elsewhere (John Beatty at Indiana U and Paul Thompson at the U of Toronto) – I had no PhD students of my own until the 1990s. Since then about ten people have worked and graduated with me – I am proud to say that they all have jobs.

Pretty crappy jobs for the most part, mark you – teaching eight courses a year on some vile prairie in the Midwest. If I had to start out again, there is no way I would be a professor – all of those years of courses and doing the marking for others, and then the stress of job hunting. I would want something that pays mega cash and yields fast cars and exotic holidays and pretty girls in my twenties, even if in my thirties I have to contemplate my misspent youth from my prison cell. I would settle for being an attendant at the Metropolitan Opera, if I could watch the performances.

Although I had no doctoral students, I taught huge classes of undergraduates – and loved every moment of it, and got several mistresses and two wives out of it. (I had a lot of catching up to do in that direction.) But, I didn't want to do nothing outside my own writing, and so in the early 1980s I founded, and for fifteen years edited, a journal in the philosophy of biology. I think it is fair to say that, from the first, *Biology and Philosophy* has been a success and it is still a success eight years after I relinquished the editorship. (I think normally it is a good thing if an editor changes after ten years – you need new ideas and visions. I kept going for fifteen years because as founder I felt I had a special obligation.)

So that was one thing I could do for my group. I am proud of the way that I (and my assistant editors and my readers and referees) worked hard with young people, helping them to improve their work and then, when it was ready, getting them an audience. In a related fashion, in the early 1990s I started the Cambridge University Press Series in the Philosophy of Biology. That has been a combination of collected essays by senior scholars and monographs by younger ones, and I think has helped people to develop and share their ideas.

I have also done a lot of editing of collections on and around the history and philosophy of evolutionary biology. A collection on the philosophy of biology for Oxford University Press with David Hull in 1998, *Readings in the Philosophy of Biology*; another collection with Hull for Cambridge University Press, *The Cambridge Companion to the Philosophy of Biology* (2007); a go-it-alone collection for Oxford, *The Oxford Handbook to the Philosophy of Biology*

(2008); a history-of-recent-paleobiology collection for the University of Chicago Press with young historian David Sepkoski, *The Paleobiological Revolution* (2009); another collection on the history of twentieth-century evolutionary theory for the American Philosophical Society with historian Joseph Cain, *Evolution in the Twentieth Century* (2009); a collection done with senior historian of science Robert J. Richards to celebrate the 200th anniversary of Charles Darwin's birth and the 150th anniversary of the publication of the *Origin of Species*, *The Cambridge Companion to the 'Origin of Species'* (2008); and now finally a massive companion to evolution for Harvard University Press with my biologist colleague Joseph Travis, *Evolution: The First Four Billion Years* (2009). What I can tell you is that I am certainly not doing any of this for the money. I just got the half-yearly print-out from Oxford for the Hull-Ruse volume, now ten years old. Although they have sold 5000 copies, they claim that we still owe them money for permissions – we are paying this off at 2.5% royalties per book.

The other thing at the social level that I have done is to engage myself in live issues in the public domain. Most notably I have fought against American biblical literalism, Creationism. I was a witness for the American Civil Liberties Union in Arkansas in 1981 (along with people like Stephen Jay Gould) and immodestly I think my spelling out of what I take to be science – the need to be lawlike, to be testable, and so forth – had great influence on the judge. It was at the heart of his ruling that Creationism is religion and not science and cannot constitutionally be taught in state-funded schools.

Many of my fellow philosophers were very critical of my engagement – they argued that you cannot demarcate science from non-science that neatly – but I feel more convinced than ever that I was right and it was a good thing that I did. In 2005 there was another trial, this time centering on so-called Intelligent Design Theory, in Dover, Pennsylvania. The philosopher for the ACLU at that event was Robert Pennock. After the Arkansas trial I brought out a collection, *But is it Science? The Philosophical Question in the Creation-Evolution Struggle*, published by Prometheus Books, out of Buffalo, N.Y. I have updated this, now with Pennock as a co-editor, and this appeared in 2008.

At the more intellectual level, I see myself as having done two important things and I am now engaged in a third. In the area of philosophy I have long pushed a naturalistic program, based on Darwinian evolutionary theory. This came out of my interest in sociobiology, especially as it applies to humans. As I have said many times, starting with my *Taking Darwin Seriously: A Naturalistic Approach to Philosophy* (1986), down through a short book on *Charles Darwin* (published in 2008 by Blackwell, with that title) for philosophers and a revision of a book for students *The Evolution Wars* (first published in 2000 and revised and republished in 2009), to my new collection *Philosophy After Darwin: Classic and Contemporary Readings* (2009), it really has to matter that, rather than the miraculous creation of a good god on the Sixth Day, we are the end products of a long slow process fueled by natural selection. It has to matter in epistemology – What can I know? – and it has to matter in ethics – What should I do?

I have always argued that the way in which we think and behave is a function of the selective process and that hence we should think of things as adaptive. We add and reason as we do because those of our would-be ancestors who did so did (survived and reproduced) better than those of our would-be ancestors that did not. Those who saw twigs broken, footsteps taken, and growls heard and who said, "Tigers, scram," did better than those who said, "Tigers, just a hypothesis not a fact." We behave as we do, again because would-be ancestors who did so did better than would-be ancestors who did not. Those who helped others tended to get help in return, and those who did not did not.

The science is catching up with the philosophy now in a big way. People like Leda Cosmides and John Tooby have done sterling work showing how it is that we think as we do for evolutionary reasons. Likewise in the field of behaviour, people like Marc Hauser are linking everything to our genes as chosen and sorted by natural selection. They are also showing that it is not just a question of behaviour but of ethics, senses of moral responsibility. We do what we do because we think we ought to do what we do – not simply because we reason that it is in our interests.

Although I was not the only one feeling my way in this direction, I am particularly pleased with my advocacy of what is often called "ethical skepticism," meaning the denial that there are foundations to moral claims. Of course this in itself is nothing new. The Emotivists, like other non-cognitivists, were in the foundation-denying business. But, as a student, Emotivism always struck me as being wrong to the point of being immoral. It is simply not the case that "Killing is wrong," means only "I don't like killing and neither should you," no matter how much hand-waving is going on. It means that killing is really and truly wrong.

The evolutionary approach helps you to make sense of the position – because if we thought that ethics was just subjective, without foundations, it would break down. So our evolution has made us think that ethics is objective – that it does have foundations – even though it doesn't. Of course this sort of thing is all in David Hume, but I am pleased at the way in which I have been able to show that a Darwinian overview makes the empirical case for the philosophy embedded within it.

In the history of science, my big book is *Monad to Man: The Concept of Progress in Evolutionary Biology*. Around 1985, with a new wife, with good health, with lots of leave, I decided that I really needed to get stuck into a major project. I wanted a long-term challenge; one that I could look back on in my old age and feel was an accomplished major feat of scholarship. I did not want to be like Terry in *On the Waterfront*, complaining that "I could've been a contender" if only I had taken the time and the effort and so forth. I got stuck into a full-scale analysis of the concept of progress as it occurs in evolutionary thinking – things getting better as they go from the blob to the human.

It turned out to be a bigger project than even I had anticipated, and although I did not always work full-time on it, in the end took some ten years to complete. (Normally, I do a book in a year, give or take.) I have to say that I was – I still am – pretty pleased with it. I have used the findings of *Monad* for another ten years, as I have completed spin-off projects. (More on these in a moment.) What I did find out was something I had not anticipated at all when I started and was a total proof that not everything is fiction, that the facts really can count.

I did not set out just to write a history of a concept. As I said earlier, I am the kind of historian of ideas who likes to use history to look at philosophical problems. Around 1980 the big question was that of objectivity/subjectivity. Is science a reflection of reality, or at least an attempt at such a reflection? This would be the position of someone like Karl Popper – science is "knowledge without a knower" – as well as my logical empiricist teachers. Or is science a creation of the scientist, immersed in his or her culture? Is science, as the followers of Michel Foucault and others said, a "social construction"? As much a part of culture as a preacher's sermon or a politician's stump speech?

I chose the idea of progress in evolutionary thought as a kind of test case. Following Kuhn in this respect, I was a naturalist in methodology as well as (as I was explaining earlier in this section) in using science as the foundation of my philosophy. I wanted to take an episode in the history of science – an episode that I as a historian of science would have to uncover – as my empirical datum to test hypotheses about the nature of science, objective or subjective.

The attractive thing about biological progress was that there was already a popular hypothesis. Philosopher Ernan McMullin particularly had made an argument that seemed to give the constructivists something while at the same time defending the castle of objectivity. He agreed fully that in the early stages of a science it is often very much a construction, reflecting the values of the society in which it is formed. An example (my example actually) would be nineteenth-century anthropology, done as it was by missionaries and civil servants and off-duty soldiers in places like India and Africa. This was full of social values about the superiority of the white man and so forth. But then, argued McMullin, the "epistemic values" kick in. These are values, like prizing consistency and simplicity and predictive fertility, that scientists adopt because they seem to lead to an understanding of reality. Over time, the epistemic values push out the cultural values. No one today could go on about the superiority of the white man because it would simply be inconsistent with what we know of modern genetics, and this latter in turn is a strongly predictive (and related epistemic values) science.

Even more attractive from my perspective was the fact that Mary Hesse had already applied something like McMullin's thesis to the concept of progress in evolutionary biology. Back at the start, in the 18th century, evolution had virtually been an epiphenomenon of progress – people believed in evolution because they took it in a progressive fashion – blob to human – and this was confirmation of their beliefs about cultural and social progress. Today nobody in biology believes in progress. The late Stephen Jay Gould called it a noxious, outdated idea which we don't need.

But why is progress now unfashionable in biology? Two reasons. First, in 1859 Darwin came along with natural selection, the survival of the fittest. This epistemically attractive mechanism – good for making predictions and so forth – is anti-progress. Which is the fittest? It all depends. Sometimes it is the intelligent and human-like. But not always. Sometimes it is the unintelligent and virus-like. Natural selection relativizes everything. Second in 1900 came the rediscovery of Mendelian genetics. The raw stuff of evolution, the building blocks, the mutations in Mendelian terms, are random, not in the sense of being uncaused but in the sense of having no direction. A new coat color could be black or green or whatever. Nothing points anywhere. Again, we have an epistemically attractive move – prediction and so forth made possible – and again we get the downplaying of the cultural notion of progress.

I however had another hypothesis. I suspected that there might be more to constructivism than these latter-day logical empiricists supposed. Could it not be that progress is no longer talked about in biology because no one – including evolutionists – believes today in social or cultural progress? Given global warming and world poverty and AIDS (remember, these were the 1980s) and much more, who dare think that things are getting better or that they ever could get better? So I was thinking that the change from progress to non-progress in evolution was a function of changed values and that the constructivists were right all along.

Well, the great thing is that neither of these hypotheses was right! I spent time in the archives of the great twentieth-century evolutionists at the American Philosophical Society in Philadelphia. I discovered that they were obsessed with their status. In the 1940s and thereabout, the time when Mendelian genetics had been fully integrated with natural selection, the evolutionists of the day – Dobzhansky, Mayr and above all Simpson – were fighting to be taken seriously as scientists. Which led to the question of why they were not already being taken seriously, which led back to the days of the *Origin* and its aftermath.

Cutting the story short – it is spelled out in great detail in *Monad to Man* – I found that there were three periods to the history of evolutionary theory. In the first, from the 18th century to the time of the *Origin*, 1859, evolution was a *pseudo science* – it was an epiphenomenon of cultural progress and no one took it seriously as science. Then after the *Origin*, the status of evolution was upgraded, but still no one thought of it as a professional functioning science. It was rather a *popular science*, a kind of secular religion, that Darwin's supporters like Thomas Henry Huxley pushed as a Christianity substitute. And selection notwithstanding, everyone interpreted evolution in a progressive fashion. Only after the 1930s and the Darwin-Mendel synthesis did evolution achieve the full status of *professional science*.

But there was a cost. All of the great evolutionists, the men just mentioned as well as others like Julian Huxley (T. H. H.'s grandson) in England got into evolution because they were attracted to the progressivism! They realized however that in order to achieve professional standing they had to remove so open a cultural concept from their science. So they did, giving us the science of today. In other words, an external-cultural value (external to science) was removed to satisfy an internal-cultural value (internal to science) – progress went in order to achieve the status of good-quality science – more accurately, to achieve the status of perceived good-quality science. My analysis not only yielded a view of the overall history of evolutionary biology that no one hitherto had seen – from pseudo science to popular science, from popular science to professional science – it threw light on the objective-subjective debate, suggesting at the very least that the tools we had been using hitherto were too crude to express the full story. Cultural and epistemic are far more mixed up than realized.

I have mentioned already that I mined *Monad* for more books over the next ten years (roughly 1996 to 2005). First came *Mystery of Mysteries: Is Evolution a Social Construction?* (1999) This went more deeply into topics about which I have just been writing, including some discussion of the role of metaphor in science. Then came *Darwin and Design: Does Evolution have a Purpose?* (2003) This did for short-term teleology – thinking in terms of ends rather than proximate causes – what *Monad* had done for long-term teleology. Finally came *The Evolution-Creation Struggle* (2005), which set out as a short synthesis of these previous books but which evolved into a discussion of how evolution has functioned (and still does function) as secular religion and argued that much of the bitterness between evolutionists and biblical literalists comes about because they are fighting for the same souls, and using the same kinds of arguments.

This last book points to what I call my present engagement in an important intellectual issue. It is in fact one that began in studying evolutionary biology but now takes me beyond. I have mentioned above my Christian background. Although this is something now far in the past, and I have no more religious belief than Richard Dawkins (to take the most extreme case of which I can think), it is still alive in that I have a real feeling of affection and gratitude to the Religious Society of Friends (to give the Quakers their full name) for the love that I received as a child – from my parents and from their co-religionists. I am sure also that my philosophical nature comes in major part from all of this, because we Junior Young Friends (as we were then called) were strongly encouraged to think for ourselves, especially about matters of morality and social concern.

I know therefore that even though evolution has never been a religion substitute for me, because of my background the entangling of evolution with religion has been a source of great intellectual (almost emotional) interest – most obviously in the fight against Creationists, but also in the history of the idea itself. A good portion of the *Darwinian Revolution* is given (in a not particularly original fashion) to exploring the debt of evolutionary thinking to Christianity. Indeed, I like to say that Darwinism is the bastard offspring of Christianity – growing out of it and in reaction to it, but every now and then in the right light looking so much like it that the heart trembles.

One thing therefore is that even though I don't believe the claims of Christianity I don't hate it. And I like many people who are Christians, despite – in major respects because of – their beliefs. After the Arkansas trial I got to know many people of faith, especially in North America, and especially those interested in reconciling science and religion. I count my connections with these people, particularly through the group with the name The Institute on Religion in an Age of Science (a rather pretentious name for a very non-pretentious group of people), as one of the real joys of my life, especially the annual meetings held on one of the Isles of Shoals, a group of former fishing islands ten miles off the coast of New Hampshire.

As I explained above, as a Quaker I never once thought there might be a clash between science and religion. But of course people like Richard Dawkins and Dan Dennett think there is and they have held forth on the subject loudly – and very successfully too I might add, given that the *God Delusion* has sold over a million and a half copies. Stuff like this is a red rag to a bull as far as I am concerned and so early in the new millennium I wrote a book, *Can a Darwinian be a Christian? The relationship between Science and Religion*, arguing that one can be both. It is not easy, I added, and then rather pompously asked "but when were the important things in life ever easy?"

The Evolution-Creation Struggle was a kind of historical counterpart to the philosophical *Can a Darwinian be a Christian?* Now I am writing a third book to make the trilogy. This is broader in being about science generally – although certainly it includes much on evolution and ties in with some earlier thoughts about metaphor. I try to show that science is successful precisely because of the limits on the kinds of questions that it asks. Beyond these limits, as scientists we have no right to say that the issues are meaningless. Hence, if religion wants to speak to issues in this realm it has every right to do so. I take my sub-title from Kant and from my religious friends – *Science and Spirituality: Making Room for Faith in the Age of Science*. Although I am dealing with science generally, I don't know enough about religion to deal with religion generally. In any case, I am inclined to think that the science-religion debate is truly a science-Christianity debate with other religions only brought in a secondary fashion. I have never been impressed with talk about Buddhism and quantum theory and that sort of thing.

I hope you can see that I am a lucky man. I love the life of the mind and I feel that I am more creative now than I have ever been. Many years ago I quit smoking so I could have more of the life I enjoy so much. Probably I should quit drinking too, but there are limits. It is all a question of balances.

3. What if any, practical and/or social-political and/or moral obligations follow from your work on evolution?

I am always hesitant about suggesting practical and social and moral implications of anything that I do. It is all a little bit too much like writing grant applications. How does my project lead to a cure for cancer and that sort of thing? So let me say, that it is the fun of working with great ideas that motivates me – that and getting out and meeting lots of people and traveling and writing books and seeing my name in print. I like the last mentioned a lot and have links on my computer to amazon.com and various of my books. I was really chuffed a year or two back when I had an article in *Playboy*; although I bought a copy in Memphis airport and was so embarrassed that I went into a lavatory cubicle to open it and look inside. Thank God there were no U.S. senators in the next cubicle. I must confess that I don't much care for all of the shaving that seems to be *de rigueur* these days.

In a way, I am a bit uncomfortable with praise – I have never been much of a person for hero worship, not even of Charles Darwin. That was part of my problem with Christianity. I cannot worship someone, especially not if they claim to be human. But I do like to be taken seriously and to have a feeling that others see that I am using my talents – that incidentally has always been the most important parable of Jesus, in my opinion. The most important miracle of course was the marriage at Cana.

But you will know if you have been reading what I have just written I do take the moral questions very seriously indeed. I try not to be preachy about these things, but I do. For me, not in any particular order, I would say that it is very important to let people know about the idea of evolution and how professional researchers have expanded and developed this idea. This is why Joe Travis and I are doing a companion to evolution. It is a terrific idea and the research is simply staggering. Fact after discovered fact makes sense only in the light of evolution – the two-hundred-million-year-old reptile discoveries in India, Antarctica, and South America, coming to life thanks to our understanding of plate tectonics; the similarity between the DNA of humans and fruitflies, making sense thanks to evo-devo (evolutionary development); the fact that worker ants are always female, a point grasped because we now understand how animals can help themselves by helping others; and more and more and more. It is a moral obligation to pass this sort of stuff on to others.

It is relatedly important to fight Creationism and to uphold evolution. We may be modified monkeys but we are a little lower than the angels, and our ability to work out the past is the best proof of this. If we are indeed made in the image of God, then this means using our intellect fearlessly, to find out truths about the world and ourselves, even when – especially when – they are uncomfortable. I should say also – I have said at length in my books, especially *The Evolution-Creation Struggle* – that the fight with Creationism is not just about the fossil record and the rest of the scientific case. It is also a philosophical fight about whether we are going to be ruled by reason and evidence – by the philosophy of the Enlightenment – or by pokey, little, cramping superstitions invented in the nineteenth century by American evangelicals. If the latter, then goodbye to women's rights and to the possibility that gays might lead full and decent and open lives like the rest of us and to the possibilities that modern medicine can solve diseases like Parkinson's and so forth. If this is not a moral fight, then I do not know what is.

At a more technical level, I do see some interesting and important work coming out of the evolutionary approach to ethics. I don't mean this in the old-fashioned sense of Social Darwinism, letting the forces of nature rule because they are the forces of nature. I am with Thomas Henry Huxley in thinking this a very bad idea, and have just prepared a new edition of Huxley's great essay, *Evolution and Ethics*, because I feel strongly about this. I should add that I say this notwithstanding the fact that my good friend Ed Wilson does think that morality emerges from the obligation to support the forces of evolution and, because he links this to our human need of biodiversity, he is doing sterling work on such topics as saving the environment, specifically the rain forests of Brazil.

Where I see interesting points for new understanding is in the exploration of the psychology of choice and moral decision making. People like Hauser are doing the empirical work, especially on such paradoxes as the trolley problem – why are we prepared to do things in the abstract but when faced with concrete problems often do the exact opposite? (The trolley problem posits two situations. One where through throwing a lever you can deflect a runaway trolley so it only kills one person rather than five. One where you do the same by pushing your neighbor onto the track. Most people are prepared to do the former but not the latter.)

For myself, I am interested in such issues as overpopulation. There are good biological reasons for having babies and equally good biological reasons for letting others have babies. If you let them have babies they will let you have babies. We have moral sentiments to back this up and this is why the Chinese policy of enforced one-child families makes us feel so uncomfortable. But perhaps the long-term prospects for our planet require us to go against the short-term sentiments, even when these sentiments are backed by moral urges or emotions. I think only by understanding the biology of the whole situation are we going to be able to disentangle different and conflicting emotions and perhaps move forward. Frankly, I don't know if any of this is true but it does seem to me to be worth exploring and it does seem to be a very important matter.

There you are. Three reasons for the moral worth of evolution and the sort of work that I do and can do. But as I said, I myself am not in this business because I am a nice person. I am in it because it fascinates me.

4. What do you see as the most interesting criticism against your position in the biological or the philosophical discussion of evolution?

I don't see any interesting criticisms against my various positions, so that is the end of that.

Of course, there are people with whom I disagree. The attempt to understand human nature in the light of our evolutionary biology is still very much a minority position in philosophy. Book after tedious book is churned out showing that it won't work. The vitriol that I encountered in the 1980s when I declared for human sociobiology was overwhelming. One critic after another went after me, declaring that I was stupid, boorish, racist, sexist, the lot. I got my own back on my 50th birthday. I made up a collage of all of their reviews, threw in some really rather nasty comments about my *Darwinian Revolution*, and in the middle of the page invited guests to come and celebrate 50 years of unbroken success. I then sent it to my critics. Philip Kitcher had the grace to smile and wish me well. Michael Ghiselin did not.

It is amazing how long, in even the most ostensibly secular of thinkers, there persist remnants of the Christian belief that we humans were made as organisms apart from the rest of Creation. But frankly, the arguments strike me as bad and, in general, perversely ignorant of the pertinent science. I should say that I am even less impressed by those who would supplement or replace basic Darwinian science with "self organization" or such things. People who believe that the unaided laws of physics and chemistry, without natural selection, can produce organized complexity – "order for free" – have been spending too much time in front of the computer screen, running over-heated programs.

In the area of the history of evolutionary theory, my good friend (and co-editor of the *Cambridge Companion to the 'Origin of Species'*) Bob Richards and I disagree about Darwin and the extent to which he was influenced by German thinking. Bob thinks him very much in the Romantic tradition; I see him as quintessentially English. We have had a lot of fun arguing this, but I am not moved one bit by his ideas. He is wrong.

Peter Bowler would take issue with my overall picture of evolution's history. He thinks that after the *Origin* there was a lot of good science, working out paths of evolution (phylogenies) and the like. He believes that this was all essential groundwork for the synthesis of Darwin and Mendel that came in the 1930s. He too is wrong, although without quite the style of Bob. The evidence is just not there. Digging up fossils in the 1870s and putting them in museums, showing that these poor dumb brutes went extinct but that we clever apes survived, was an important part of the social story. It has nothing to do with the Hardy-Weinberg law or the other foundations of modern evolutionary biology.

Bowler's problem is that he spent so much of his life working on these second-rate scientists of the late nineteenth and early twentieth century that he is convinced that they must be important – if they are not, then he has been wasting his time. Historically, however, they are important but only because they show us what was not happening. At a time when physics was about to make its greatest breakthroughs ever, the evolutionists were putting on museum displays for immigrant kids, to show them that they too could strive to be like real white men if they worked hard and copied their masters. I am right – evolution around 1900 was pop science, secular religion. No amount of work on the chief actors of the time suggests otherwise.

As far as the science and religion interface is concerned, frankly my main concern is with being associated with a field where so much simply dreadful stuff passes for scholarship. On the side of the religion friendly, too often the aim is to get something for nothing. People want to derive God from the empirical world, a trick that Darwin in the *Origin* stopped once and for all. You just cannot do it, but nevertheless there are all sorts of neo-vitalists going around today trying to show that "order for free" and that sort of thing tells us that there is SOMETHING MORE. The latest nonsense is the so-called "anthropic principle" that apparently shows that the constants in the universe are so finely tuned, in the sense that they make life possible, that their existence and nature cannot be pure chance. Think of a number, double it, and the answer you want is half. How anyone thinks they can say that these constants are uniquely chosen beats me.

On the side of science friendly, or more precisely religion hostile, what passes for argumentation makes me sweat. As I have said about the *God Delusion*, it makes me ashamed to be an atheist. It is the first time in my life that I felt sorry for the ontological argument. (The argument that proves God's existence from his nature.) Of course in part this is a turf war. It is okay for me to run down philosophical ideas. It is quite another when outsiders do. But my personal irritations do not excuse really rotten arguments based on ignorance of the true positions of those being critiqued. What depresses me is that that particular book is so badly written, by a man whose *Selfish Gene* still strikes me as one of the most elegant books ever written. There is simply nothing to be learnt from books like that or from related works like Dennett's *Breaking the Spell* that claims to prove that religion is a parasite like the lancet fluke. Bad, bad, bad.

5. With respect to present and future inquiry, how can the most important problems concerning evolutionary theory (or evolutionary aspects of your field) be identified and explored?

Philosophers are the worst of all people to make predictions. We live in the past, have trouble with the present, and should leave the future alone. My personal feeling is that my generation creamed the interesting problems. When people like Hull and I got involved, in the mid to late 1960s, the philosophy of biology was moribund. We had terrific fun getting it up and going. For myself, by about 1980 I was losing interest in the more technical stuff. To spice up my life, I got involved in the sociobiology debate – as I just said above, I got my money's worth there. In any case, as I have also explained, my interests became much more those of traditional epistemology and ethics and how evolutionary theory might throw light upon them.

Analogously, when I got interested in the history of evolutionary theory, in the early 1970s, there was still enough original stuff to be dug out. I remember the thrill of finding that one of the Darwin notebooks had been mistranscribed and the related thrill of finding a pamphlet that Darwin had read that contained an early intimation of natural selection, that he had noted with marginalia. (I am not suggesting that Darwin cribbed the idea. The author had no intention of making anything of the point.) Then it was possible, first for me to put together what we all knew in an overall survey, followed in the succeeding years with the massive study of progress – a study that I think led me to get the big picture when no one else had it. (I was lucky not being a full-time historian of science, especially not one setting out. Big pictures are not encouraged. Find some obscure corner and ferret away is the prescription – not a bad one to begin but not one that should rule the rest of your life, as it often does.)

I feel a bit the same way about the science-religion field. I think I have got an insight that is going to change or rather create the standard against which other work will be judged. Immodest perhaps, but it cannot be worse than what exists already.

So that is how I feel, but now let me take it all back, or at least put it into doubt. If, at the end of 1953, Watson and Crick had thought that from now on it is all going to be downhill – you cannot discover the double helix a second time – I think they might have had a point. And yet we know that the last fifty plus years have been incredibly exciting for molecular biology and there is so much work still to be done. Why should this not be the case in the areas where I work? Obviously, as the science of evolutionary biology develops, there will be new or transformed questions for philosophy. This holds across the spectrum, but particularly in the realm of human biology. Truly, for all of my optimism and partisanship, we have very far to go making out the roots of biology in human thought and behavior.

I suspect that that is also going to be much work for history, and its practitioners. Apart from anything else, so much of the work so far has been so directed to the Anglophone world. I expect that there is much more to be learnt about Europe – in Russia particularly. Also, too few historians of science really know much about the societies in which their scientists lived. At the moment, regular historians still look down on historians of science and so there is not the interchange of ideas one needs – historians of science tend to be isolated even when they are in history departments. (They mix with scientists and others on campus who are more user friendly.)

And as far as science and religion is concerned, as I have said, thus far we have hardly started. Even if I succeed beyond my dreams, I am only one person and there is masses amount of work for others. I really would be interested to see if, in a hundred years, the science-religion field exists at all. There are days, candidly, when I think the whole enterprise is doomed, kept alive only by the massive funding of the Templeton Foundation – and now that the old man is gone, I wonder how long that will last. In recent years, the big prize has gone to folk in the science-religion field. Look for that to start going in the direction of right-wing, Jesus freaks. Perhaps it will go to the first person to breed a flawless red heifer, so vital for the Temple sacrifices that Revelation leads us to expect in the near future now that the Jews have returned to Israel.

Now, my questions answered, I will close. Eight years ago (in 2000), after 38 years, I left Canada and moved down to Florida. I did so simply because then in Ontario there were compulsory retirement laws, and I would have been forced out at 65. America has no such rules and I can work until I drop, which I intend to. But until then, lubricated with gallons of cheap Australian red wine, I am having more fun than I have ever had before. I have great friends in the field – David Hull (never a cross word in forty years of friendship), Bob Richards (a deeply moral man and a brilliant scholar), the leading historian of science and religion Ron Numbers (that is one area where the science-religion field is thriving and deservedly so), and many others, including younger ones like the historians Joe Cain and David Sepkoski – and I have great younger colleagues and students here at Florida State

University. Darwinian evolutionary theory has been very good to me and it is a privilege to try to make some return.

19
Geerat J. Vermeij

Department of Geology
University of California at Davis, USA

An Evolutionary Worldview

1. From natural history to evolution

A passion for natural history has been with me since early childhood. I was introduced to nature by my parents and older brother Arie in the polders of the Netherlands, where I spent the first nine years of my life. It was a world of abundant wildflowers, rustling poplars along the dikes, alders and willows and waterlogged meadows along muddy ditches, and the pervasive smell of sweet grass mixed with cow manure. My family, as well as my early teachers at boarding schools for the blind I began attending when I was not quite four years old, encouraged me to explore and to observe nature even if that meant the occasional encounter with a thorny rose or blackberry bush or with the nettles that grew so prolifically in the tall grass. I began collecting all sorts of natural objects — pinecones, acorns, beechnuts, and even pebbles — and on those special occasions when we bicycled from Gouda to the beach at Scheveningen, I spent my time sifting the sand for shells.

At the institute for the blind at Huizen, nature provided a welcome if solitary haven from a generally unhappy existence. I was drawn not only to the pine woods on the school grounds, but also to the institute's fine collection of stuffed birds and mammals. The teachers were uniformly outstanding, and children could progress at their own pace. They strongly emphasized the skills of careful observation, and enriched the necessary tedium of language and arithmetic with much more appealing subjects like science, geography, and Dutch history. It was also at Huizen where I read my first book on fossils. The book told of the recovery of the skeleton of an ancient lizard-like animal — a mosasaur, or Maas lizard — from what I now know to be the classic Late Cretaceous locality at the Sint Pietersberg near Maastricht. I do not recall any sense of strangeness about the discovery of animal remains millions of years old. Like the natural history of which I was so fond, these bones simply intrigued me. They fed my unbounded but still unfocussed curiosity about the natural world.

When my family and I arrived in New Jersey in 1955 after I had just celebrated my ninth birthday, I was confronted with a wholly new natural order, one of an unruly wilderness with tall trees, poison ivy, dramatically noisy snow crickets at nights and summer cicadas by day, and wood thrushes singing songs of unexcelled beauty and complexity. I was fascinated by the enormous contrast between this new, somewhat forbidding nature and the much more benign realm of life I had left behind. Contemplating this difference, I found my love of nature gradually growing into a passion for science.

This transformation received a huge boost from Mrs. Colberg, my fourth-grade teacher. On her annual sojourn to the Gulf coast of Florida, she picked up shells that she then brought back to her classroom, where I instantly became intrigued with them. These shells were so strikingly varied and beautiful, so utterly foreign to my previous experience with Dutch shells, that I resolved at once to become a conchologist. With my command of English now good enough to read books, I absorbed what I could about science, shells, marine life, plants and animals, fossils, the stars and planets, rocks and minerals, and more. My father and Arie possessed a talent for making accurate raised illustrations, enabling me to become familiar with maps as well as with the shapes of objects I could not observe firsthand. With their help, and with my mother's hand-copying books into Braille and her reading aloud in both Dutch and English, I embarked on a frenzied course of learning that continues to this day.

My first exposure to evolution came at age eleven, when I read a book about the scientific accomplishments of Lamarck, Darwin, and other great naturalists of the past. From the beginning, evolution made sense to me. Although I had been drawn to Christianity as a lonely and unhappy boarder at the school for the blind in Huizen, I already had doubts about God and about the stories as told in the Bible before I came to the United States. Religion therefore never impeded my straight-line path toward science in general and the study of evolution in particular. General principles always appealed to me, and evolution — even the very primitive and incomplete version to which I was exposed in the late 1950s

— offered a pleasing and rational way to organize and think about the facts of natural history as well as a coherent account of Earth's past.

During my student years at Princeton and Yale, my diverse interests in shells, geography, and science came together, unifying and expanding as I deepened my understanding of evolution and its productions. There was never any doubt about my goal: I wanted to become a research scientist. With the arrogance and optimism of youth, I hoped to contribute not only to the study of shell-bearing animals, but also to the broader aspects of evolutionary theory and the history of life.

As a graduate student and later as a young professor, I approached evolutionary science from the safe confines of ecology, paleontology, taxonomy, biogeography, and the Neo-Darwinian synthesis. Field work on molluscs carried me to all parts of the world, and provided me with a wealth of comparisons. Later, as I turned more and more to fossils, I applied these insights from living nature to the strange worlds of the geological past. The shapes of fossil shells provided clues about conditions of the past as perceived by the shell-builders themselves, and these conditions were often quite different from those prevailing today. The view of history and evolution that I found most agreeable was one in which living things change according to circumstances that could be either directly observed or inferred from fossils and their enclosing sediments. This was evolution in an ecological and functional context.

2. Arms races and economics

My studies led me to the realization that the history of life consists of episodes of escalation, a sort of arms race in which competition among top consumers or top producers in an ecosystem created intense selection in favor of increasingly sophisticated defenses and other means to retain or acquire locally scarce, contested resources. Disturbances like mass extinctions temporarily put an end to, or occasionally even reversed, these episodes, but escalation resumed when these constraints were lifted. Times and places of abundant resources, especially periods and regions where warmth and prolific nutrients allowed life to explore a wide variety of adaptive possibilities, were most favorable to the process of escalation and to the evolutionary introduction of energy-intensive innovations, which enabled their possessors to extract and hold resources more effectively.

The arms races I was documenting bore a striking resemblance to the military arms race between the United States and the Soviet Union, prompting me to wonder if insights into evolutionary escalation could be applied to questions about how arms races start and end in the human realm. I began to read some of the relevant literature in political science and economics, and soon came to realize that evolution and economics were closely allied expressions of fundamental processes — competition, cooperation, trade, synergy, feedback, and adaptation — that pervade human relationships as well as the domain of nonhuman life. Evolution was no longer just an organizing theory for biologists and paleontologists; together with some basic economics, it is the foundation for understanding history. It provides a unified, coherent theory accounting for the phenomenology of life in all its extraordinary variety and complexity, past and present.

There was, of course, much more to evolution than the Neodarwinian synthesis. Organisms do not simply respond to their living and inanimate surroundings; they also create and construct their environment, an environment of diversity, complexity, interdependence, and unintentional regulation at scales ranging from the ephemeral and local to the timeless and biospheric. And all this evolutionary phenomenology, and the historical patterns to which it gives rise, flows from the universal competition between individual living entities for locally scarce resources. Enabling factors — resources and the factors controlling their abundance and supply — provide the evolutionary potential, or the extent to which living things can respond effectively to opportunities, allies, and enemies; the intensity of competition ultimately determines how much of this potential is realized. Resources influence organisms, and organisms influence (and often cooperate to stabilize or enhance the supply of) resources. The most power-intensive members of a system exercise a disproportionate influence over other organisms as well as over the distribution of resources. Relentless competition between these dominants imparts an arrow of time — toward greater power and reach — not only among the dominant competitors but among successive ecosystems. These fundamental relationships among metabolizing entities apply as much to advanced human civilizations as they do to ecosystems of reefs, rain forests, the deep sea, energy-starved caves, and the biosphere as a whole.

Evolutionary aspects of economics, and economic aspects of evolution, have been widely discussed ever since Darwin set out his theory of evolution in 1859. Most scholars, however, have considered the similarities between evolution and human-economic life to reflect analogy rather than common cause. I contend instead that the similarities spring from processes and principles that are intrinsic to life. There is a single, fundamentally economic theory of competition-driven dynamics between living things and the resources that are essential to survival and propagation. Moreover — and from my perspective most interestingly — the theory provides a scientific basis of history in both the human realm and the domain of nonhuman life.

This economic view of evolution, combining patterns of genealogy and replacement with organic structure and function, differs in important respects from other, currently more fashionable, conceptions. Like Darwin's original theory, my evolutionary worldview requires inheritance of adaptive traits, but it is not as gene-centered as are more other versions of the theory. Adaptations do, of course, spread through genetic inheritance, involving transmission vertically from generation to generation, but there is abundant evidence as well for horizontal spread of genes, for symbioses creating novel organisms, for cultural evolution, for epigenetic effects on gene expression that are environmental in origin. Besides its gene-centered orientation, the Neodarwinian synthesis achieved in the middle of the twentieth century emphasized selection over all other agencies of adaptive evolution. In my conception, the all-important process of selection is augmented by effects that the organism has on its environment. It is selection together with this environmental construction by living things that yields the universally appreciated good fit between organism and environment.

The macroevolutionary tradition, which gained popularity in the 1970s and continues to hold sway among paleontologists today, is concerned with the evolutionary comings and goings of units above the level of the organism, that is, with species, genera, families, and clades (branches in the evolutionary tree of life, each consisting of an ancestor and all of its descendants). Its proponents maintain that properties of these larger units — geographic ranges of species, population size, collective ecological distribution, and other group traits — affect, and indeed are often more important in the long run than, the traits and selective regimes of individual organisms. These collective traits determine the susceptibility of clades to species formation and extinction. However, given that the characteristics of individuals influence the collective properties of more inclusive categories, macroevolutionary patterns cannot in my view be divorced from, or be considered independent of, the ecological (or economic) processes that exercise selection among individuals or among cooperative groups of individuals (coalitions and societies). Moreover, clades and even species are not monolithic units; they are internally heterogeneous in habitat, body size, style of defense, mode of feeding, reproductive type, and other features. When new opportunities arise for clade expansion, as when members colonize islands or invade larger regions not previously occupied by them, members often acquire novel traits that materially affect macroevolutionary behavior. Cichlid fish in African lakes have speciated wildly, whereas in tropical America their diversification has been modest. Evolution — and the properties of clades and species — is opportunistic, incorporating change as well as conserving traits.

None of these other worldviews — strictly gene-based inheritance, the Neodarwinian synthesis, and macroevolution — fully embraces what I perceive to be the scope of evolutionary theory. Moreover, these incomplete theories do not take full account of what I take to be life's fundamental distinguishing characteristics. These are (1) metabolism, which has the inevitable effect of modifying an organism's surroundings and which provides life's power to grow, propagate, and adapt; and (2) interaction, which determines who wins a contested resource and how that resource is distributed among organisms in the larger economy those lifeforms inhabit and create. An economic view of evolution applies also to the growth and development within individuals. The unity of process in the history of life and the history of humanity is far more obvious when adaptation and fate are causally linked to economic performance and resource supply than when theory is expressed in more narrowly drawn terms.

3. Applications

As I see it, the economic evolutionary perspective incorporates laws that are every bit as powerful as are the laws of physics and chemistry. There will, for example, always be competition, and in almost all competitive interactions the outcome is unequal for the parties concerned: one party will gain more (or lose less) resource than the other. Greater power (work, or energy, per unit time) is linked to disproportionate influence over the phenotypes, activities, and distribution of organisms, and over the rate and distribution of resource supply. Some adaptive traits — cooperation, increased metabolic rate, mobility, and the ability to respond rapidly to a wide range of unpredictable circumstances — are universally beneficial and will be favored except during short-term disruptions. These rules, which are based on first principles as well as on abundant empirical evidence, are unlikely to be broken no matter how technologically sophisticated the organism or its constructions are. Utopian schemes that would seek to eliminate competition in favor of cooperation, or that propound the notion that human economic growth no longer depends on resources, are therefore bound to fail. Economic evolutionary theory and its implications for history therefore provide coherent predictions about what is possible and what is not. Together with the realization that no adaptation can in principle be 100% effective, regardless of its degree of sophistication or specialization, this distinction between the possible and the impossible is the most important lesson we can draw from an inclusive evolutionary worldview.

There are, of course, many other specific implications as well. One of the most intractable problems facing human civilization is how to cope with unpredictable phenomena such as terrorist acts, unforeseen weather-related catastrophes, and the spread of new diseases whose pathogenic agents are unknown to our immune systems. The evolutionary history of life can be mined for strategies that work in coping with rare events, which in principle cannot be predicted or, in the language of science, cannot be incorporated into an adaptive hypothesis of the environment. All organisms living today have, after all, descended from ancestors whose lineages had to endure at least five — and likely more — crises leading to mass extinctions. We know, for example, that the ability to shut down metabolism for a protracted time was key to surviving these biotic cataclysms, and that this ability likely arose as an adaptation to the much more mundane — indeed almost quotidian — events such as predation, sudden heat or cold, unexpected submergence in water for air-breathing animals, and the like. The applicability of everyday adaptations, which were incorporated into the adaptive hypotheses of individuals, to rare events represents one way in which the unpredictable is made predictable. Another is the formation of organic architecture that is flexible, redundant, and rapidly responsive. These attributes, too, are the products of true adaptation, because they confer economic advantages to their bearers in everyday economic interactions. Incorporation of these and other ideas into policies related to security, public health, and climate change has the potential to make those policies both more effective and more realistic. We must work with nature, not against it, and we must take advantage of three and a half billion years of experience that organisms have collectively accumulated.

Another vexing issue for mankind is the tendency for civilizations to overexploit resources and therefore to become more vulnerable to the inevitable disruptions stemming from climatic variation. Humans are by no means the only species to live unsustainably. Evidence indicates that forest trees collectively deplete the soil of nutrients, and that they become stunted and less productive without replenishment from outside. Traditionally, human civilizations have staved off such collapse either by managing resources sustainably — a rare occurrence, and one possible only in highly productive settings — or by engaging in trade, effectively taking advantage of subsidies from elsewhere. The latter solution entails greater global interdependence, which works as long as resources are being produced in surplus in at least some locations. But such systems survive only if source populations and source reservoirs continue to exist, a possibility enhanced when such sources are multiple. Highly interdependent economies and individuals hold an advantage during good times, but are vulnerable to collapse when exposed to disruptions in production. To avoid such collapse, significant production should be widely distributed, and not concentrated in just one region or by just one or a few suppliers. By emphasizing the long-range consequences of short-term economic circumstances, the economic-evolutionary perspective should force policy-makers to be wary of market-driven economic advantages that tend to apply only over the short run and on local or regional rather than global scales. In short, economic policies could substantially improve with a better understanding of evolution.

4. Criticisms

An evolutionary theory based on economic interactions has certainly not been universally welcomed. In part, it represents a departure from the prevailing trend, begun with the birth of theoretical ecology in the 1950s, and flourishing with the rise of anti-adaptationism and the macroevolutionary perspective in the 1970s and 1980s, toward abstraction; and in part it raises doubts among both economists and evolutionists that the complexities of human-economic life can be reduced to mere variants on economic themes established at the dawn of life three and a half billion years ago. As a lover of natural history, I find it well-nigh impossible to treat organisms, the traits of organisms, or communities of organisms, as abstract entities. Even a casual observer will quickly realize that, as in human communities, the world of organisms is one replete with risks and rewards, competitive and cooperative relationships, and limitations and opportunities. Even if not everything about an organism is an adaptation in the struggle for survival and propagation, it seems inescapable that organisms are well adapted to the conditions at hand, and that such adaptation is enforced in part by repeated bouts of competition, predation, and other forms of resource redistribution. Yet there arose in the 1970s a backlash, spearheaded by Stephen J. Gould and Richard C. Lewontin, against what was dubbed the "adaptationist programme", the idea that adaptation was the assumed rather than a demonstrated condition of life. Gould, Lewontin, and their numerous uncritical followers considered many — perhaps even most — traits as by-products of the way organisms are built. Furthermore, they regarded adaptations as functioning under very particular — and therefore ephemeral — circumstances, and that such events as the great biotic crises had a far greater influence on organic architecture and the composition of communities than did adaptation. As a result, it became unfashionable for paleontologists and many other scientists to think about organisms as adapted beings. Adaptation existed, of course, but its role in evolution was overblown and uncritically accepted, according to Gould and Lewontin's influential arguments.

Although this line of reasoning had the beneficial effect of making adaptive inferences more rigorous, the antiadaptationist rhetoric of scholars as prominent as Lewontin and Gould did far more harm than good. It diverted attention from the evolutionary implications for ecology, made the study of function in plants and animals a dead-end enterprise that was said to contribute little of value to evolutionary theory, misled countless people about the nature and broad applicability of adaptation, distorted one of the most important aspects — the link between performance and fate — of evolutionary change, and needlessly isolated the study of evolution from such outside influences as economics and the genetics of behavioral traits that would enrich it.

One aspect of my view of adaptation and escalation that has been seen as controversial is the idea that adaptation is most likely under conditions of population expansion. Traditionally, competition and selection in favor of a new phenotype are thought to be strongest when a population is under stress, that is, when it is either stable or in decline. It follows from this line of thinking that competition and selection are relaxed when resources are abundant enough to allow populations to grow. In my view, this argument fails to differentiate between phenomena at the population level and interaction among individuals. A given resource — mates, food, space — may be abundant for the population as a whole, allowing the population to expand; but it may be locally scarce, and therefore the target of intense competition and selection, for individuals within the population. In a growing population or economy, individuals can respond to intense competition adaptively by altering patterns of allocation to the many conflicting functions the individual must perform. Moreover, novel traits that are beneficial but which may come with small disadvantages can persist in a growing population and be improved through subsequent selection. In a static or declining population, such innovations would likely be eliminated quickly because the collateral disadvantages of the new trait pose too large an opportunity cost. Competition and selection are likely always to be intense, but the ability to respond adaptively to them is, in my view, largely limited to conditions in which economic growth is permissive, leaving room for error and for a much longer apprenticeship of new traits.

A widespread belief held by biologists and human-oriented scholars alike is that the human species has surpassed other life-forms by such a wide margin that the principles governing nonhuman organisms and ecosystems have been superseded by phenomena specific to a highly intelligent, articulate, literate, intentional, moral, conscious, cooperative species that has discovered wholly new principles of energy extraction, communication, and economics. Humans have forged all manner of institutions unknown in the realm of nonhuman life — religions, schools, governments, banks, and more — and engage in trade and knowledge-based innovation on a scale beyond anything seen in nature. Humans rely on machinery and energy sources that other forms of life do not use. Much of our power, in other words, derives from extrasomal extensions of our bodies.

There can be no question that humans, individually and collectively, have gained more power than any other species has acquired. As a result, we can do things on vastly larger spatial scales and vastly shorter time scales. We have expanded (though not fundamentally altered) the ways in which information and adaptation spread. But what we have not done, and are unlikely to do, is to violate fundamental economic laws, those governing competition, tradeoffs, feedbacks, synergies, and the distribution of resources.

To deny the biological nature of humans or the economic nature of nonhuman life-forms is to ignore or fail to recognize important common principles that have a high potential to enrich both evolutionary biology and the human-social sciences. In the long run, assigning a unique status to objects or phenomena — the sun, the Earth, the United States, God — is a fundamentally unscientific act, because unique things resist either experimental or comparative inquiry. Uniqueness gives license to pursue policies or engage in practices that fly in the face of scientific reality and discard well-supported theories about how the world works.

5. How do we proceed?

Despite my strong attraction to general principles and the development of an inclusive, coherent theory of evolution, I am at heart an explorer. By closely observing natural history — ecology, morphology, behavior, and the physical environment — in both the present and the geological past, and by reading as deeply and widely as possible, I am in search of puzzles, of objects and phenomena that do not make sense. Much as I greatly appreciate mathematical and experimental approaches, my preferred modes of inquiry are comparative and synthetic. I notice differences and similarities, and apply knowledge from many sources to construct and test hypotheses to uncover patterns, and to place the new insights into an evolving explanatory theory.

It has become fashionable in educational circles to disparage rote learning and to emphasize "creativity". But creativity without knowledge or disciplined thinking is chaos. It is impossible to make progress in science without a firm command of a great deal of knowledge — facts, ideas, conundrums — and without applying rigorous yet flexible methods of inquiry. It is impossible to think about things without facts and experiences. And it is impossible — or at least fraught with perils — to introduce new ways of thinking about old problems without a willingness to confront established convention and authority. To teach science, therefore, we must emphasize not only the methods and technology of inquiry, but also the ability to observe, to ask questions, and to be in command of a large body of knowledge.

What, then, do we need to learn about evolution, and how do we proceed? At the top of my scientific agenda is to make history a fashionable, and I would argue an essential, style of doing science. Despite the ascendancy of molecular phylogenetics, biology in general and evolutionary biology in particular remain shockingly ahistorical disciplines. Many practitioners in these and other fields of science continue to cling to the idea that experimental science is the only legitimate science. They are unaware of, or simply dismiss as hopelessly inadequate, major strides in historical science made by paleobiologists, anthropologists, geologists, and even astronomers. Not only is history a legitimate subject for empirical inquiry, founded on comparative data and multiple lines of evidence, but a theory of history should be an integral part of a general theory of evolution. I have already pleaded the case for the role of economic principles in such a theory, but there is clearly also a crucial role for the incorporation of principles governing development, genetic architecture, gene expression and its controls, scaling laws, biomechanics, and the physiological control of complex living systems. Evolutionary theory is, in other words, a theory of function as well as of structure. And both function and structure have a history.

A second, likely more contentious, aim is to woo theoretical biology away from physics and toward a more holistic conception in which the behavior and properties of groups — populations, ecosystems, coalitions, and societies — must be added to those of more traditional units of evolutionary interest, such as individuals and genes. Where physicists might express biological phenomena in such terms as information, entropy, and complexity, I would substitute language that is more directly relevant to the living and fates of living things. This substitution does not apply only to interactions and adaptations, but also to structural and organizational aspects of organic form. In the case of structure, the language of engineering might be more suitable than, say, the abstract terminology of theoretical thermodynamics.

Third, we desperately need a better theory and methodology for inferring ancestor-descendant relationships in the evolutionary tree of life. Despite a vast and increasingly impenetrable literature on the procedures and statistical analyses of phylogenetic inference, far too little attention has been paid to the assumptions underlying these methods. Many of these assumptions are highly questionable. Why, for example, should evolution be parsimonious, that is, proceed with the fewest possible steps from the base to the tips of the tree? Why must all branching be dichotomous? Why should the distribution of characters in outgroups affect inferences about the branching pattern in the ingroup if evolutionary transitions are independent in all branches of the tree? In my view, evolution is an inherently opportunistic process, for which global methods of phylogenetic inference are unsuitable. Instead of relying on computer programs with largely hidden universal assumptions, we should construct phylogenies piece by piece, much as human genealogists construct family histories. This involves careful attention to the ecological and functional plausibility of proposed evolutionary transformations, a deep understanding of the development and function of both morphological and molecular traits, serious consideration of the available fossil record, and continual appraisal of assumptions and procedures. Furthermore, although the reconstruction of phylogenetic trees is clearly essential for answering a host of questions, we must not ignore such phenomena as horizontal gene transfer, anastomosing branches, symbiosis between phylogenetically distant organisms into single evolutionary units, and clade replacement as important contributions to observed historical patterns.

We shall need every approach at our disposal for answering outstanding questions. Among these questions are: (1) How does the emergent property of consciousness evolve? (2) Does evolutionary theory imply the existence of large-scale trends or predictable patterns of change, as I have come to believe, or is it consistent with a pattern of dominant contingency, in which the particulars of initial conditions and pathways are the chief determinants of evolutionary change? (3) Is the rate of adaptation observed in nature now or in the past fast enough for most species to respond effectively to the rapid and in some cases unprecedented changes — habitat fragmentation, overexploitation of resources, pollution, and alterations in the physical nature of the atmosphere and ocean — that humans are bringing about?

Finally, we must do our utmost to overcome the public's fear and distrust of the phenomenon and theory of evolution. This will, I think, involve firstly an effort to acquaint people from an early age with the beauty that remains in nature, and to awaken in them the values and pleasures of wondering and questioning. Secondly, we must confront directly the widespread belief that science in general, and evolution in particular, robs life of its purpose and meaning. In my teaching, I approach this topic by arguing that purpose and meaning derive from one's own efforts and from a deep appreciation and understanding of ourselves, our relationships, and our fellow life-forms. It is we, not some unknowable supernatural entity, who are responsible for making our lives meaningful. Given the social-adaptive nature of mythology and ideology as means to strengthen group cohesion in human society, and the universal existence of fear of the unknown, I am not optimistic that humanity can construct a moral social contract of individual freedom and the common good on a rational basis without the unquestioning allegiance to deities or their human surrogates, but we must try.

20
Andreas Wagner

Professor, Department of Biochemistry, University of Zurich

The University of Zurich, The Santa Fe Institute, The Swiss Institute of Bioinformatics

Switzerland

1. What are the most important problems in evolutionary theory?

The most important problems regard the evolutionary origins of innovation and the origins of cooperation. The first problem is very close to my research. A relatively small community of researchers work on it, perhaps because it is a very difficult problem. The key question is how biological systems are able to produce novel things, evolutionary innovations. This is one of the most important unsolved problems in (evolutionary) biology. Darwin's theory has essentially left it untouched. Although Darwin was aware of its importance, his theory was fundamentally unable to explain how genuine novelty can arise from the minor variants that natural selection acts upon. The importance of this question becomes clear by considering the political dimension it has taken in intelligent design (ID) creationism. This pseudoscientific current views evolutionary innovations as beyond the grasp of natural explanations. As I write these lines, its advocates have been quite successful in changing high school curricula in the United States to include teachings that evolutionary biology may be fundamentally unable to understand evolutionary innovation. I call ID pseudoscientific, because its fundamental tenet implies that we should stop trying to understand how evolutionary innovations arose.

In my group's work, we take the ability of biological systems to evolve innovations as a given, and ask a more specific question: Do living things have any special properties that allow them to innovate? Are these properties different from the principles on which systems engineered by humans are built? In this work, we focus on molecular systems, molecules such as proteins and RNA, regulatory systems such as transcription factor networks, and metabolic networks. There are two main reasons. First, ultimately, most evolutionary innovations can be understood as a sequence of small molecular changes in these systems. Second, to understand innovation, we need to understand the relationship between genotype (the genetic material of an organism) and phenotype (any observable feature of the organism). This understanding is easiest to come by in molecular systems. What we have learned from studying such systems is quite striking. Biological systems have an amazing internal flexibility or robustness. They can dramatically change their interior make-up, their genotype, while preserving their functional properties and phenotype. This robustness is crucial for their ability to evolve a wide variety of new features, features that are ultimately responsible for life's success on this planet.

Cooperation, the second paramount problem, I will discuss only briefly. Others in this volume will undoubtedly say much more about it. The origin of cooperation is important to all areas of biology. Take the evolution of multicellularity. How did cells, unthinking lumps of protoplasm "learn" to sacrifice life itself? Without this sacrifice, multicellular life would never have evolved. It arose more than a billion years ago, for reasons that we still do not fully understand. What we do understand, however, is that the cooperation of cells in the interest of a greater good has been important to life's phenomenal success. Another example involves societies, from insect colonies to human organizations. To understand how cooperation arises in them is key to understand their fabric, how conflicts are resolved in them, why conflicts sometime fail to be resolved, and how cooperation can be promoted. When applied to human societies, the problem also acquires a moral component.

2. What does your work reveal about biological evolution that other academics, citizens, philosophers, or biologists typically fail to appreciate?

My work touches upon many questions in evolutionary biology. I will merely speak to one class of problems that has occupied my group recently, the problem of evolutionary innovation. Let me first draw an analogy to politics, where we can distinguish two broad political philosophies: Conservatism, which attempts to preserve time-tested ways of living; and liberalism, which is more open to the new, and to changing the *status quo*. Liberalism and conservatism are opposite extremes of a spectrum, and on any one issue, it may be impossible to be both a liberal and conservative.

An analogous dichotomy can be found in the evolution of organisms, human or otherwise. Their features or phenotypes are often extremely well adapted to the world around them. However, this phenotype is constantly perturbed, either through mutations in the genotype, or through changes in the internal or external environment of the organism. We learned in recent years that phenotypes can be highly robust to such changes (e.g., Wagner, *Robustness and Evolvability in Living Systems*, Princeton University Press, 2005). But at the same time, it may be necessary for organisms to change evolutionarily and to innovate. Some such innovation may require entire new ways of making a living, such as the ability to use new food sources, or the ability to catalyze new chemical reactions. How do organisms preserve what works in the face of mutation, while at the same time being able to innovate? The axis liberalism-conservatism in the political analogy above corresponds to the axis robustness-innovation in the biological realm. It may seem that you cannot do both: You cannot be good at preserving the status quo, while being an innovator. But appearances can deceive. We and others have shown that organisms may accomplish an amazing feat. They may have found a way to avoid this dichotomy. Some systems we study show that the greater their robustness to change is, the greater is their ability to produce evolutionary innovation. For example, we have demonstrated this property for enzymes (e.g., Ferrada and Wagner, Proc. Roy. Soc. London Ser. B, 2008). We examined many different protein structures that differ in their robustness to genetic (amino acid) change, and analyzed the diversity of enzymatic reactions catalyzed by enzymes with a given structure. Such functional diversity is a past record of evolutionary innovation. We showed that proteins with highly robust structures – the phenotypes of proteins – have experienced more functional innovation in their evolutionary history then less robust proteins. It thus seems that phenotypic robustness facilitates evolutionary innovation. Others, including Jesse Bloom and Dan Tawfik, have demonstrated a similar phenomenon in the laboratory. They showed experimentally that protein folds more tolerant against amino acid changes also evolve new enzymatic functions more readily. Briefly, the reason is that highly robust systems can have many alternative genotypes with the same phenotype. This increases the likelihood that a mutation of any one such genotype produces a new phenotype that is the solution to a biological problem (in this case a problem of chemical catalysis). In the submicroscopic world of molecules and molecular networks, "conservatism" and "liberalism" are no contradictions.

3. What practical implications follow from your work on evolution?

Human engineers need to pay even more attention to how biological systems innovate than they do now. The usage of evolutionary principles in engineering is of course not new. Protein engineers mutagenize large populations of proteins in order to produce new functions. Metabolic engineers subject microbial populations to specific nutrient environments in order to evolve new metabolic abilities. And computer scientists employ evolutionary principles to solve difficult optimization problems or optimal design challenges. However, especially computational approaches are still too far from biological reality to mimic the success of biological systems. The key issue is how to represent genotypes and phenotypes computationally: In living systems from molecules to networks, genotypes relate to phenotypes in specific ways that facilitate innovation. For example, the sets of genotypes with the same phenotype are typically very large, and random changes in mutations of two different genotypes that have the same phenotype may yield completely different novel phenotypes. These two ingredients are important for the ability of biological systems to innovate, and our current engineering approaches take insufficient advantages of it.

4. What do you see as the most interesting criticism against your position in the biological or philosophical discussion of evolution?

The first point to highlight is not so much a criticism of my work, but the curious indifference of many biological researchers – in particular biomedical researchers – to evolution in general. We cannot make sense of the origins of disease, of genome architecture, of the workings of organisms, and of virtually any aspect of biological systems without taking an evolutionary perspective. The "why" is key to the "what" in biology. Yet many researchers have only the most naïve and rudimentary understanding of evolution. More generally, I have always been very curious about why different kinds of questions attract different individuals. Many of the scientific questions that might keep me awake at night would leave some of my molecular biology colleagues completely cold, and vice versa. What drives one's scientific interests? The likely cause is a complex mix of personal pre-disposition, access to good senior scientific role models, and social reward systems. This mix and its consequences might be worth studying further.

An issue closer to my personal research agenda is "hard selectionism". It is a view based on the (accurate) observation that most genetic variants occurring in populations affect the fitness of organisms at some point in their history. Their fate is thus determined by natural selection. From this view, a hard selectionist would conclude that neutral mutations, mutations that affect fitness not at all or too little to be visible to selection, do not matter in the evolutionary process. Hard selectionism is an old and established view, and perhaps the dominant view among evolutionary geneticists. It can be contrasted with neutralism, the view – largely refuted by whole-genome data – that most such variation does not affect fitness. Neutralism and selectionism have broad implications on our understanding of how evolutionary innovations arise.

Molecular engineering and evolutionary genetics work that focuses on phenotypes (molecular structures and functions, gene expression patterns, and metabolic abilities, to name a few) demonstrates a much more subtle relationship between neutral and other mutations. Specifically, recent experimental and computational work shows that robust biological systems – from molecules to networks – that can tolerate many genetic changes without losing their phenotype, have a leg up in producing evolutionary innovations, new phenotypes with new functions beneficial to the organism. Any given genetic change is more likely to be neutral in such robust systems than in less robust systems. However, neutral mutations may only be neutral shortly after they arise, and may become affected by selection later-on, after other genetic changes arise, or in a new and different environment. From this point of view, neutral mutations pave the way for adaptation, and may become beneficial (or deleterious) only later. The role of neutral mutations is analogous to the role exaptations play in morphological evolution.

Selectionism contains an interesting but subtle flaw that is symptomatic of how we like to view natural phenomena, namely in *essentialistic* terms. This means that we seem to have innate difficulties in viewing properties of biological phenomena as depending on the context in which they occur (although we pay much lip-service to such context-dependence). When applied to selectionism, this means that we have difficulty appreciating that a mutation that was neutral at some point in time may become either beneficial or deleterious later on, depending on other genetic or environmental changes. (Problems with essentialism are not new in biology. The problem of how to define biological species properly is another example of how our thinking can be plagued by it.) If, however, we can accept this notion, then we can accept the importance of neutral mutations for evolution, while at the same time embracing the importance of natural selection in explaining genetic variation.

5. Why were you initially drawn to research in evolutionary biology?

I still remember the day in school when we learned that many organisms are diploid, that they have two copies of each gene. I remember asking my high school teacher whether diploidy evolved because it protected against "failure" of genes. He was a competent teacher whose classes where rigorous and engaging, but this question baffled him. He just said evasively that the matter might not be quite that simple. (Unbeknownst to both of us he was right, but I would not learn that until a decade later.) I soon forgot about this question, and it would not re-emerge until many years later when I became interested in why biological systems – from genes to whole organisms – are robust to genetic change.

On a different occasion, our high school teacher introduced the concept of homeostasis in physiology. Our didactic introduction to homeostasis was not a biological system, mind you, but a flushing toilet. This rather lowbrow example illustrates the powerful yet simple principle that feedback regulation can use a sensor (a float connected to a valve) to bring a controlled variable (the water level in the toilet tank) back to a set point (the fully filled tank) after a perturbation (the act of flushing). Since my first exposure to it, I was fascinated by this abstract language, because it is able to subsume a great variety of biological phenomena into a single, generalizing principle. The teacher deserves a great amount of credit for exposing high school biology students to such material.

These two are the earliest occurrences I recall of themes that would later play an important role in my work: Questions about ultimate (evolutionary) causes rather than proximate (mechanistic) causes, and general principles that unify a diverse range of phenomena.

After high school, at the university level, I would train primarily in molecular biology. I was never an avid naturalist, interested in cataloguing and systematizing natural history. Fortunately for me, mine was among the first generations of biologists that could hope to have a career without being naturalists. Molecular biology attracted me, because it seemed to offer explanations about why organisms function they way they do. However, because molecular biology was about mechanistic explanations for cellular processes, it did not satisfy my taste for evolutionary questions. I also eventually tired of absorbing the innumerable details of complicated molecular process, for example, of the complicated networks comprising dozens of proteins that communicate information from the cell surface to DNA. What had seemed like a molecular explanation of biological processes to a beginning student, became a mere description once I had become sufficiently immersed in the literature. It did not help that none of my advisors in molecular biology encouraged me to pursue the questions I was interested in. I thus soon turned away from molecular and to evolutionary biology.

My interest in ultimate causes, combined with a taste for general principles, explains much of the direction my research has taken. To give one example, in recent years my laboratory has been studying the evolution of very different classes of systems, RNA and protein molecules, regulatory gene networks, and metabolic networks. The intent of this work is to identify deep and non-obvious commonalities between these systems that could explain how they produce evolutionary innovations. This is an ambitious goal, and it is not certain that we will be successful (but we will certainly be unsuccessful if we do not try.)

In sum, to me, evolutionary biology is not so much about cataloguing the diversity of life, or about explaining our place in it. Although I greatly respect those who dedicate their life to these problems, I was always drawn to the most general principles of biological organization, those which may hold for all of life. Evolutionary biology has been my access road to them.

21
David Sloan Wilson

SUNY Distinguished Professor

Departments of Biology and Anthropology

Binghamton University, USA

---

1. Why were you initially drawn to discussions and research on evolution (or evolutionary aspects of your field)?

I was lucky to be present at a time when the fields of Ecology, Evolution, and Behavior were merging into a single discipline with evolutionary theory as the theoretical foundation. During the first half of the 20th century, ecology and behavior were largely descriptive sciences, often organized by taxonomy (e.g., ornithology, entomology, botany) or habitat (e.g., tropical biology, marine biology, limnology). Most ecologists did not yet appreciate the degree to which all organisms can be understood in terms of the same basic evolutionary principles, and most evolutionary biologists were preoccupied with questions such as the mechanisms of heredity that did not take them into the field.

When I entered the University of Rochester in the late 1960's, I had decided to become a scientist to distance myself from my novelist father (as I recount in my book *Evolution for Everyone*) but I had a white-coated laboratory scientist in mind, perhaps a brain surgeon or someone who discovers a cure for cancer. I frankly found that kind of science too hard and discovered another kind of science—ecology—that allowed me to spend time outdoors and study whole organisms, which I already loved to do. By chance, I started studying zooplankton and entered graduate school at Michigan State University intending to be an aquatic ecologist.

The graduate program at Michigan State was small but dynamic. Earl Werner and Don Hall were using optimal foraging theory to predict the feeding behavior of fish and the consequences for the structure of ecological communities. Based on the assumption that fish are adapted by natural selection to maximize the amount of energy harvested per unit time, they predicted optimal feeding behavior on the basis of mathematical models. The predictions were not obvious; no one had previously guessed that foragers should rank their food according to energy divided by handling time and accept less preferred items only to the degree that more preferred items are not available! The behavior of real fish in laboratory experiments did not exactly correspond to the predictions but came impressively close. Even better, optimal foraging theory did a good job of explaining habitat use, habitat switching, and the coexistence of fish species in artificial ponds and natural lakes. Better still, the same theory transcended taxonomic boundaries, explaining the foraging behavior of lizards, insects, and birds just as well as fish. Optimization theory even transcended *subject* boundaries, for example by predicting optimal *mate* choice as easily as optimal *food* choice.

Using evolutionary theory to predict the properties of organisms in relation to each other and their environments turned me from an aquatic ecologist to an *everything* ecologist and behaviorist. I could easily switch from studying vertical migration in zooplankton to the feeding behavior of zooplankton (a subject switch), to the feeding behavior of ant lions (a taxonomic shift). Even though I showed no aptitude for math as an undergraduate student, I now had a strong incentive to build mathematical models. My first theoretical paper was titled "The Adequacy of Body Size as a Niche Difference" (Wilson 1975a) and showed that a difference in body size allows species to coexist only under special circumstances; in other cases bigger species outcompete smaller species. It eventually became a citation classic. Then I started modeling a completely different subject—group selection—which determined the future direction of my entire career (Wilson 1975b).

My point is not to boast about personal accomplishments but to illustrate a *style of reasoning* that is based upon evolutionary theory and provides a common language for studying all subjects and organisms. This integration was in progress during the 1970's and I was lucky to be present.

2. What does your work reveal about biological evolution (or evolutionary aspects of your field) that other academics, citizens, philosophers or biologists typically fail to appreciate?

I have two major answers to question 2. First, the integration described in my answer to question 1, is still in progress within the biological sciences. For example, a major insight of evolutionary theory is that every trait that evolves requires two explanations; one based on the physical mechanisms that cause the trait to exist (proximate causation) and the other based on the historical and environmental forces that cause the trait to exist, compared to many other traits that *could* have existed (ultimate causation). Mature evolutionary research programs pay equal attention to both proximate and ultimate causation in ways that are mutually informing. However, it is still common for some biologists to rely primarily on ultimate causation (e.g., why moths evolve to match their background without any attention to the genetic, development, and physiology of moth coloration) and others to rely primarily on proximate causation (e.g. the genetic expression of coloration without any attention to the evolutionary "big picture.")

The study of nature at large spatial scales (e.g., ecosystem and landscape ecology) is also poorly integrated with evolutionary theory. Keeping track of energy and nutrient flows lumps the actions of many species and creates broad categories such as "autotroph," "herbivore," "detritivore," and so on. This conceptualization ignores the fact that each of these broad categories includes myriad of species that are evolving to increase their survival and reproduction. Their adaptive strategies can have a powerful effect on the higher-level ecosystem processes and therefore cannot be ignored. For example, beavers are optimal foragers that preferentially eat trees that are low in tannins. Trees low in tannins are competitively superior to trees high in tannins, creating a trade-off between herbivore defense and competitive ability. As a result, enormous individual differences exist within a single species of trees such as poplars. When beavers move into an area, they selectively remove the low-tannin trees. The leaves shed by the remaining high-tannin trees decompose much more slowly and change the chemistry of the water, changing the faunal composition of aquatic community and ecosystem processes. Thomas Whitham (e.g., Whitham et al. 2008) has coined the term "ecosystem genetics" to underscore the fact that genetic evolution takes place on ecological time scales and cannot be ignored in the study of ecosystem processes.

Early models of community and ecosystem dynamics made the naïve assumption that species influence each other primarily through their densities. For example, if two competing species, A and B, share a common predator (C), then adding C should reduce competition between A and B by reducing their densities. What could be simpler? Yet, in real experiments involving two species of competing sunfish and their bass predator (Turner and Mittelbach 1990), in the absence of bass the sunfish safely utilized the entire pond. When bass were added, the sunfish quickly sought refuge in the vegetated littoral zone, where they competed much more than before. The bass had virtually no effect on the *numbers* of sunfish, but the effect on their *behavior* cascaded throughout the entire aquatic ecosystem. It is not an exaggeration to say that the study of nature at large spatial and temporal scales cannot be understood without studying the component species as rapidly evolving entities employing sophisticated adaptive strategies.

My second answer to question 2 concerns the debate over multilevel selection theory. Darwin was perceptive enough to distinguish between individual-level and group-level adaptations. Individual-level adaptations cause some individuals to survive and reproduce better than other individuals within their immediate vicinity. Examples include moths that are more cryptic, hawks with better eyesight, and polar bears with thicker fur. Individual-level adaptations are *locally advantageous*. In contrast, consider social adaptations such as warning others about predators, refraining from overexploiting one's resources, or helping others in distress. These solid citizen traits might be "for the good of the group," but often they are *not locally advantageous*. Indeed, they are vulnerable to exploitation by traits that we associate with free-riding, cheating, and exploitation.

How can a socially advantageous trait evolve when it is locally disadvantageous? Darwin reasoned that these traits could be positively selected at a larger scale; groups of solid citizens can outcompete groups of slackers, even if slackers outcompete solid citizens within groups. This was the birth of multilevel selection theory, which arguably has the most turbulent history of any major concept in evolutionary theory. Group selection was falsely rejected in the 1960's and decades were required to restore the simple logic of Darwin's reasoning to mainstream evolutionary theory. E.O. Wilson and I have recently reviewed the current status of multilevel selection theory (Wilson and Wilson 2007, 2008), which need not be repeated here. In addition, I have discussed the degree to which the group selection controversy represents a failure of the scientific process in a series of blogs titled "Truth and Reconciliation for Group Selection"[1].

3. What, if any, practical and/or social-political and/or moral obligations follow from your work on evolution?

In my answers to the previous questions, I have discussed the integration taking place within the biological sciences, which were sufficiently advanced that by 1973 Theodosius Dobzhansky could utter his famous statement that "nothing in biology makes sense except in the light of evolution." In contrast, the study of human-related subjects from an evolutionary perspective has lagged behind the biological sciences. Certain subjects such as physical anthropology and human genetics are based on evolution, of course, but the closer one gets to human behavior and culture, the more evolutionary theory is avoided. The publication of E.O. Wilson's Sociobiology (1975) provides an example. Wilson's encyclopedic review showed that the social behavior of all nonhuman species, from ants to primates, could be understood in terms of the same evolutionary principles, just as Dobzhansky said. Yet, Wilson's final chapter on humans created a firestorm of controversy.

It wasn't until the 1990's that evolutionary science began to expand beyond the biological sciences to include all human-related subjects, like water bursting from a dam. I gravitated toward this movement early, in part because it resembles the novelistic enterprise of trying to understand the human condition that I knew through my father. One of my earliest publications describes a cultural route to biological fitness (Wilson 1978). My interest in group selection was easy to apply to the subject of human morality, just as Darwin had done in *Descent of Man*. Pursuing this line of inquiry led to my books *Unto Others: the evolution and psychology of unselfish behavior* (Sober and Wilson 1998) *Darwin's Cathedral: Evolution, Religion, and the Nature of Society* (Wilson 2002), and *Evolution for Everyone: How Darwin's Theory Can Change the Way We Think About Our Lives* (Wilson 2007).

Studying all human related subjects from an evolutionary perspective is already in progress at the level of scientific research and scholarship, but it is not yet reflected in higher education, where evolution is still taught primarily as a biological subject at virtually all colleges and universities. To address this problem, I and my colleagues at Binghamton University initiated EvoS (for Evolutionary Studies, pronounced as one word) a campus-wide

---

[1] http://www.huffingtonpost.com/david-sloan-wilson/#blogger_bio

evolutionary studies program that is described in detail elsewhere (Wilson, Geher and Waldo 2009). EvoS is now in the process of expanding into a worldwide consortium of programs with the help of NSF funding, suggesting that a tipping point has been reached in the study of humanity from an evolutionary perspective (see http://evostudies.org/).

If higher education lags behind scientific research and scholarship, then using evolutionary theory to inform public policy and improve the quality of everyday life lags still farther. The most discouraging fact about public awareness of evolution is not that 50% of Americans don't believe the theory, but that nearly 100% don't connect it to matters of importance in their lives. Politicians are unable to state their own views about evolution and their advisors (outside of biology) almost certainly did not receive training during their own higher education. Organizations dedicated to promote awareness of evolution are vastly outgunned by organizations such as the Discovery Institute dedicated to spreading misinformation about evolution. To address this issue, I have become involved in two programmatic activities comparable in scope to EvoS. The first is the Binghamton Neighborhood Project (`http://evolution.binghamton.edu/bnp/`), which is designed to understand and improve the quality of life from an evolutionary perspective at the scale of an entire city (Wilson, O'Brien and Sesma 2009). The second is the Evolution Institute, the first think tank for informing public policy from an evolutionary perspective (`http://theevolutioninstitute.org`).

These are modest beginnings, but I believe that the 21st century will be viewed by historians as a period of integration for the study of humanity, comparable to the integration of the biological sciences that took place in the 20th century. Knowledge about evolution will become as essential for our wellbeing as knowledge about physics and chemistry.

4. What do you see as the most interesting criticism against your position in the biological or philosophical discussion of evolution?

Pluralism has become a hot topic in philosophical discussions of evolution, especially in relation to multilevel selection. I find this literature frustrating. Surely, there is a legitimate form of pluralism, whereby a complicated subject such as multilevel selection can be better studied from a diversity of perspectives than from a single perspective, much as a mountain can be better viewed from a number of angles than from a single angle. However, the different perspectives must come to an agreement on factual matters, just as people viewing a mountain from different angles must come to an agreement about the contours that they are viewing. If different perspectives fail to converge on factual matters, then pluralism runs a genuine risk of becoming like postmodernism in the humanities, where anything goes.

There is also a danger of too many perspectives, even when they can be related to each other. For example, there are now many theoretical perspectives for the study of social behavior, which use the same terms such as "individual selection," "altruism," and "relatedness" in subtly different ways. Every new perspective adds a cost of potential semantic confusion that needs to be justified by added insight, yet there is no formal process for conducting such a cost-benefit analysis.

In practice, pluralism often obscures the resolution of debates that should be more widely acknowledged as facts upon which everyone can agree. For example, consider the question of whether a species of algae is limited by nitrogen or phosphorus. This is not a matter of perspective but a matter of deciding among two mutually exclusive possibilities, or discovering an interesting interaction. In the same way, when Hamilton developed inclusive fitness theory, he thought that it could explain the evolution of altruism *without invoking group selection.* This was not a matter of perspective but a matter of deciding between two different processes. Later it became apparent that inclusive fitness theory assumes the existence of multiple groups, that altruism is selectively disadvantageous within groups and evolves only by virtue of groups with the most altruists differentially contributing to the total gene pool. In other words, inclusive fitness theory *does* invoke group selection and its apparent difference is merely a matter of perspective, based on a method of calculating total gene frequency change without explicitly examining gene frequency change within single groups. There might be practical value in employing inclusive fitness theory, but it cannot be regarded as a denial of group selection as a process. Yet, there is still a widespread assumption that inclusive fitness theory as an alternative *perspective* has the same significance as when it was regarded as an alternative *process.*

Another widespread criticism of evolution in relation to human behavior, including my own work, is that if evolution refers to all types of change, it explains nothing by explaining everything. My response to this is as follows: consider the standard definition of genetic evolution as all kinds of gene frequency change, whether by selection, mutation, drift, linkage disequilibrium, and so on. The definition needs to include everything so it can function as a complete accounting system. What saves the definition from being vacuous is that outcomes are assigned to meaningful subcategories on a case-by-case basis. This gene evolved by selection, that gene evolved by drift, and so on.

When we consider evolution in relation to human behavior, we need to include other categories of change, based on cultural evolution and psychological processes such as imitation and intentional thought. Just as with genetic evolution, it is important to have an accounting system that includes all kinds of change, which includes meaningful subcategories that can be determined on a case-by-case basis. This trait evolved by a raw process of cultural variation and selection, that trait by a process of unconscious imitation, that trait by a process of conscious intentional thought, and so on. In this fashion, evolutionary theory offers an integrative theoretical framework that includes previous perspectives.

Yet another widespread criticism of evolution in relation to human behavior is the specter of genetic determinism. If our behavior is determined by our genes, and if our genes can't change over time scales that matter for human welfare, then it must mean that human behavior can't change, much as we might like it to. Learning and culture are then conceptualized as alternatives to evolutionary theory.

Along with most of my evolutionary colleagues, I regard this conceptualization as profoundly mistaken. The capacity for learning and culture are both products of genetic evolution and fast-paced evolutionary processes in their own right. They are much better understood from an evolutionary perspective than as alternatives to evolutionary theory. Nevertheless, there is a sense in which evolutionary theory in its current form is guilty of genetic determinism. All of the classic models of social behavior, including group selection models, assume that behaviors are coded directly by genes. Hardly anyone regards the assumption as biologically realistic but it is still made to simplify the math. Only now are we beginning to realize that relaxing the assumption by making behavior less genetically deterministic can have profound effects on the outcome of the models.

As a second example, the school of evolutionary psychology associated with Leda Cosmides and John Tooby emphasizes specialized modules that evolve by genetic evolution and are highly selective in their environmental inputs. Open-ended learning and cultural transmission are excluded from evolutionary psychology as part of the so-called "standard social science model" (e.g., Tooby and Cosmides 1992). There is also a growing literature on open-ended cultural evolution, but it is poorly integrated with the version of evolutionary psychology outlined above. Truly avoiding the charge of genetic determinism will require a greater appreciation of open-ended processes of psychological and cultural change in our species.

5. With respect to present and future inquiry, how can the most important problems concerning evolutionary theory (or evolutionary aspects of your field) be identified and explored?

When I write about evolution for a general audience, I sometimes describe it as like a crystal ball for making informed predications about all aspects of life. It is extraordinary, when one pauses to think about it, that evolutionists can move from subject to subject and organism to organism over the course of their careers, not as amateurs, but at the highest level of scientific inquiry. It is even more extraordinary to contemplate that the same crystal ball can be used to make informed predictions about all aspects of humanity. Just as I was lucky to begin my career when the fields of Ecology, Evolution, and Behavior were merging, we are all lucky to partake in this even wider integration.

Literature Cited

Dobzhansky, T. (1973). Nothing in biology makes sense except in the light of evolution. American Biology Teacher, 35, 125-129.

Sober, E., & Wilson, D. S. (1998). Unto Others: The Evolution and Psychology of Unselfish Behavior. Cambridge, MA: Harvard University Press.

Tooby, J., & Cosmides, L. (1992). The psychological foundations of culture. In J. H. Barkow, L. Cosmides & J. Tooby (Eds.), The adapted mind: evolutionary psychology and the generation of culture (pp. 19-136). Oxford: Oxford University Press.

Turner, A. M., & Mittelbach, G. G. (1990). Predator avoidance and community structure: interactions among piscivores, plankivores, and plankton. Ecology, 71, 2241-2254.

Whitham, T. G., DiFazio, S. P., Schweitzer, J. A., Shuster, S. M., Allan, G. J., Baily, J. K., et al. (2008). Extending genomics to natural communities and ecosystems. Science, 320, 492-495.

Wilson, D. S. (1975). A theory of group selection. Proceedings of the National Academy of Sciences, 72, 143-146.

Wilson, D. S. (1975). The adequacy of body size as a niche difference. American Naturalist, 109, 769-789.

Wilson, D. S. (1978). A cultural route to biological fitness. Evolutionary theory, 3, 235-236.

Wilson, D. S. (2002). Darwin's Cathedral: Evolution, Religion and the Nature of Society. Chicago: University of Chicago Press.

Wilson, D. S. (2007). Evolution for Everyone: How Darwin's Theory Can Change the Way We Think About Our Lives. New York: Delacorte.

Wilson, D. S., Geher, G., Waldo, J., & Change, R. (2009). The EvoS Consortium: Completing the Evolutionary Synthesis in Higher Education. EvoS Journal, 1, In press.

Wilson, D. S., O'Brien, D. T., & Sesma, A. (2009). Human Prosociality from an Evolutionary Perspective: Variation and Correlations on a City-wide Scale. Evolution and Human Behavior, 30, 190-200.

Wilson, D. S., & Wilson, E. O. (2007). Rethinking the theoretical foundation of sociobiology. Quarterly Review of Biology, 82, 327-348.

Wilson, D. S., & Wilson, E. O. (2008). Evolution "For the Good of the Group". American Scientist, 96, 380-389.

Wilson, E. O. (1975). Sociobiology: the new synthesis. Cambridge, Mass: Harvard University Press.

Index

A

Adaptationist programme 30, 214
Adaptationist 118, 135
Adaptionism 82, 83, 195, 237
Adelson Glenn 42,
Agnosticism 88,
Alberch Pere 112, 154
Hall Don 229,
Altruism 21, 59, 115, 181, 236
Altruistic 59, 89, 146, 148, 157, 175, 176, 181
Anagenetic change 31
Ancestor-descendant relationship 218
Antagonistic pleiotropy 33
Anthropic principle 203
Antiadatationism(st) 214
Anti-essentialism 43, 159
Anti-preformationism 159
Anti-reductionism 105
Anti-science 14
Anti-evolution 1, 97, 99, 129
Aristotle 21, 148, 188
Atheism 88
Axiomatic view

B

Baldwin effect 154
Bateson William 1, 50, 128
Beatty John 113, 192
Beckner Morton 113
Binmore Ken 144
Bloom Jesse 224
Bonner John Tyler 130
Bowler Peter 203

Boyd Robert 84
Brandon Robert 113
Burke Edmund 182

C

Cain Joseph 192
Cambrian explosion 136
Cavalli-Sforza Luigi L 84
Cheater mutant 185
Christian (s) 14, 188, 198, 200, 202
Co-evolution 69
Cooperation 174, 176, 222
Cosmides Leda 83, 117, 195, 237
Coyne Jerry 115
Creation 29, 138, 193
Creationism 1, 26, 29, 99, 222
Creationists 93, 95, 98, 99, 100, 133, 181, 198
Crick Francis 138
Cultural evolution 22, 43, 71, 76, 84, 150, 236
Cultural norms 180
Cuvier Georges 129

D

Darden Lindley 112
Davidson Eric 130, 139
Dawkins Richard 2, 20, 40
Dennett Daniel 40, 144, 198, 203
Design 19, 21
Developmental biology 32
Developmental Systems Theory (DST) 40, 77, 152–162
Developmentalists 152

Dobzhansky Theodosius 1, 25, 119, 190, 197, 233
Dretske Fred 38
Dynamic attractors 125
Dynamical patterning modules (DPMs) 135

E

Economics 14, 144, 148, 150, 159, 209, 211, 216, 221
Ecosystem genetics 232
Eliminative reductionism
Emergence 12, 89
Emergent 16, 48, 114, 126, 145, 218
Entomology 27
Epidemiological transition 105
Epigenetic heritability 78
Epigenetic inheritance 69
Epigenetics 4, 68
Essence 41
Essentialistic terms 226
Ethical scepticism 195
Ethics 89
Evidence 29
Evo-devo 7, 149
Evolutionary approach to ethics 201
Evolutionary biology 9, 15, 27
Evolutionary epistemology 59, 61
Evolutionary genetics 26
Evolutionary physiology 33
Evolutionary psychology 30, 83
Evolutionary synthesis 30–32
Evolutionary theories 25
Evolutionary-developmental 131
Evolutionists 152
External environment 223

F

Faith 29
Fantastic bonus 120

Feldman Marc 84
Female orgasm 113, 118
Feyerabend 112
Fisher 30, 45
Fitness difference 42
Food choice 231
Formal Darwinism 149
Foucault Michel 196
Free will 21
Free-floating rationales 21
Frisch HL 130

G

Game theory 174
Geisteswissenschaften 21
Gene-centered 138
Genetic assimilation 31
Genetic determinism 236
Genetic selfishness 175
Genotype 4
Ghiselin Michael 57
Giere Ronald N. 113
Godfrey-Smith Peter 21
Goldschmidt Richard 128
Goodnight Charles 115
Goodwin Brian 126
Gould Stephen Jay 20, 29, 30, 31, 40
Grafen Alan 144
Gray Russel 161
Green revolution 105
Greene John 191
Gregorian creature 20
Grene Margerie 112
Griesemer James 113
Griffith Paul 120
Group selection 30, 146
Group-level adaptations

H

Haig David 45, 177
Haldane 103
Hamilton Bill 20, 30, 175, 236
Hard selectionism 226

Hauser Marc 1
Hempel Carl 190
Hesse Mary 196
Higher order synergies
Historicism 86
Holliday Robin 68
Homeorhesis 105
Homeostasis 108
Homosexuality 62
Hopeful monster 128
Horizontal spread of genes 211
HOX genes 9
Hull David 41, 43
Hume David 195
Hunt Gene 115
Hutton James 138
Huxley Julian 98
Huxley Thomas Henry
Hyper- selectionist 105

I

Inclusive fitness theory 236
Individual level adaptations 232
Individual selection 236
Information 32
Inheritance of aquired characteristics 66
Innovation 222
Integrated wholes 48
Intelligent design/ ID 19, 29, 93, 133
Intentional thought 236
Interactors 57
Internal environmet 223
Interspecific interaction 25

J

Jablonski David 115

K

Kafastos Fotis 130
Kauffman Stuart 48
Keller Evelyn Fox 113
Kerr Ben 40

Kin selection 30, 62, 146
Kitcher Philip 38, 112, 121, 202
Kuhn Thomas 30, 112, 190, 191, 196

L

Lamarck 129, 136, 208
Lamarckian 68, 69, 72, 103, 133, 154, 188
Lamarckism 72, 128, 129
Lamb Marion 67, 71, 77, 79
Levels of selection 19, 57, 58, 82, 83, 104, 114, 144, 145, 148
Levins Richard 136
Lewontin Richard 29, 30, 38, 40
Life cycle 6, 7
Llloyd Lisa 40
Longino Helen 121
Lyell Charles 138
Lysenko/Lysenkoism 72, 103, 129

M

MacArthur Robert 9, 25, 104
Macroevolution 95, 98
Major evolutionary transitions 145
Mate choice 231
Maturana Humberto 162
Mayr Ernst 25, 43, 53
McMullin Ernan 196
Memes 61, 164
Memetics 84
Mendel(ian) genetics 9, 72
Mendel Gregor 1
Mendelian system 45
Microevolution 98
Millikan Ruth 38
Mind-body problem 38
Modern synthesis 72, 131
Molecular biology 32
Molecular evolution 33
Molecular genetics 9
Molecular variation 30

Moral consequences 43
Moral decision making 201
Morphological evolution 30
Morphology 48
Multilevel selection theory

N

Nagel Ernst 53
Natural kinds 58
Natural selection 9, 30
Nature-nurture dualism 154
Naturwissenschaften 21
neoDarwinianism 15
neural nets 19
Neutralists 67
New synthesis 10
Nicolis Gregoire 126
Numerical taxonomy 53

O

Ontogeny 1, 7, 67
Ontological 14
Optimization 231
Organism 56
Origin-of-life question 100
Orr H. Allen 115
Orthogenesis 128, 134
Oyama Susan 40

P

Paleobiologists 26, 30, 33
Palmer Craig 117
Peirce Charles Sanders 11
Phenotype 2, 4, 30
Phenotypic evolution 31
Phylogenetic 33
Phylogeny 7, 67
Popper Karl 196
Popperian creature 20
Porter Roy 191
Postmodernist 15
Pradeu Thomas 161
Pragmatic naturalism 86
Prigogine Ilya 126

Primary replicators 59
Professional science 197
Psychology of choice 201
Punctuated equillibria 26, 30, 67
Purity as a goal 43

R

Rationalism 29
Reductionism 9, 12, 105, 144, 145
Reichenbach 121
Relatedness 236
Religion 86, 87
Replicator(s) 40, 41, 57
Resister mutant 185
Richards Robert J. 192, 202
Richerson Peter 84
Robustness 224
Rosenberg Alexander 143
Roughgarden Joan 178
Rudwick Martin 191
Ruse Michael 53, 61
Rykiel Edward 114
Ryle Gilbert 188

S

Saltationism 128
Saunders jr. John W. 130
Schaffner Ken 113
Scheffer Sonja 27
Schmalhausen 103
Science of qualities 48
Secular humanism 88
Secularism 88
Selection 56, 63
Selectionists 67
Self organisation 165
Selfish gene 2, 177
Selfishness 174
Self-organizing 14
Self-reproducing 14
Semiotic 14
Semiotic theory 12

Sepkoski David 192
Sexual dimorphism 62
Sexual selection 178
Simpson George Gaylord 53, 128
Skinnerian creatures 20
Skyrms Brian 114
Slobodkin Lawrence 25
Smith John Maynard 20, 30
Sober Elliott 38, 90
Social conditions 105
Social construction 196
Social evolution 174
Social insects 67, 175, 176, 181, 185
Social learning 71
Social norms 180
Sociocultural evolution 61
Soft inheritance 71
special creation 138
Speciation 26
Species selection 31
Spencer 146
Sterelny Kim 38, 45
Strassmann Joan 176
Symbolic communication
Sympatric speciation 27
Synthesis 32
Synthetic theory of evolution 30
System approaches 160
System biology 9
Systematics 53
Szathmáry Eörs 79

T

Tawfik Dan 224
Teleological 2
Teleology 198
Thompson D'Arcy 128
Thompson Evan 162
Thompson Paul 113

Thorhill Randy 117
Tinbergen Nico 1
Tooby John 117
Trivers Robert 20

U

Utility
Unit of development 161
Unit of selection 113, 115, 117, 120
Unit of transmission 71, 72
Valery Paul 172

V

Van Fraassen Bas 112
Van Valen Leigh 107
Varela Fracisco 162
Variation 41
Vehicle 41, 57
Vrba Elisabeth 115

W

Waddington 31, 67, 103
Wade Michael 115
Wallace Bruce 25, 107
Waters Ken 120
Watson 205
Werner Earl 229
Whitham Thomas 232
Wiesmann August 131
Williams George 20, 30, 115
Williams Mary 113
Wilson David Sloan 30, 178
Wilson Edward O. 29, 104
Wilson Robert 120
Wimsatt William 38, 112
Woodger JH 113, 121
Wright Sewall 30, 115

Y

Young JZ 19
Young Robert 191

www.ingramcontent.com/pod-product-compliance
Lightning Source LLC
Chambersburg PA
CBHW021806220426
43662CB00006B/206